Springer Textbooks in Earth Sciences, Geography and Environment

The Springer Textbooks series publishes a broad portfolio of textbooks on Earth Sciences, Geography and Environmental Science. Springer textbooks provide comprehensive introductions as well as in-depth knowledge for advanced studies. A clear, reader-friendly layout and features such as end-of-chapter summaries, work examples, exercises, and glossaries help the reader to access the subject. Springer textbooks are essential for students, researchers and applied scientists.

Jamison Conley

A Geographer's Guide
to Computing Fundamentals

Python in ArcGIS Pro

 Springer

Jamison Conley
Department of Geography and Geology
West Virginia University
Morgantown, WV, USA

ISSN 2510-1307   ISSN 2510-1315   (electronic)
Springer Textbooks in Earth Sciences, Geography and Environment
ISBN 978-3-031-08500-0   ISBN 978-3-031-08498-0   (eBook)
https://doi.org/10.1007/978-3-031-08498-0

Cover Image credit: Art by Jocelyn J. Nolan

This Springer imprint is published by the registered company Springer Nature Switzerland AG
The registered company address is: Gewerbestrasse 11, 6330 Cham, Switzerland

This book is dedicated to my family, and especially my wife, Lee Ann Nolan. Without her guidance, this book would not have begun. Without her patience and support, this book would not have been completed. I am grateful to her for all her help, and a dedication here is insufficient thanks, but it's a nice start.

Preface

Like many textbooks, this book is an outgrowth of classroom activities that I have designed over several years and iterations of teaching a course on GIS programming. It grew out of programming lab exercises that I had developed, and the recognition that having the exercises integrated with the rest of the material, instead of separated into their own sections, or online supplemental material, greatly assists students' understanding of how the computer science principles work and helps them apply those principles, especially when shifting from one programming language to another.

Like students in my classes, I am not assuming that you, the reader, have any background in computer science or programming experience. I am, though, assuming you are very familiar with ArcGIS Pro. If you are unfamiliar with ArcGIS Pro, there are many resources available from ESRI and other sources. ESRI offers a range of Web trainings through http://www.esri.com/training/, which may provide a quick way to get up to speed on ArcGIS Pro.

Here is how I envision you might use this book. I developed it for use in a course I teach on GIS programming, although you might be coming to it from the perspective of a researcher or professional who wants to employ Python in your GIS work, instead of the student perspective.

There are 21 chapters organized into 4 parts. Part I, titled Getting Acquainted with Python, which has Chaps. 1–6, serves as an introduction to computer science principles in general and Python in particular. This part is important for those with little or no programming experience, or in other words, those for whom terms like variable, statement, and function are unknown. Part II moves from general programming and Python to the application of programming within ArcGIS Pro. It does so through the lens of data structures, such as rasters and vectors. Part III builds upon data structures and then addresses the creation of algorithms and the development of new tools within ArcGIS Pro to implement and use those algorithms. Finally, Part IV expands upon this and then moves to more complex programs that involve multiple Python files, and the development of Python Packages to use these within ArcGIS Pro.

In a classroom setting, I use Parts I–III in a fast-paced single-semester course. With more time, I would suggest a two-course sequence starting with Parts I and II in the first semester and Parts III and IV in the second semester. In a non-classroom setting, I would encourage those who have little or no programming background to start with Part I, while those who have programming experience and only want to apply existing Python knowledge to ArcGIS Pro can jump into Part II and beyond.

Most chapters are structured to encourage readers to understand both the computer science principles that underly programming within a GIS context as well as the application of those principles. This comes from having the following structure, which embeds practical exercises within the text of the chapter. They begin with opening remarks, which usually introduce one or more relevant computer science concepts for the chapter in question. Then there is a series of guided tasks which take the student through the concepts introduced in the opening remarks. These tasks also elaborate on the remarks, presenting more complex aspects of the topic alongside the software within the task. The guided tasks have instructions for readers to complete within Python and ArcGIS Pro, as well as questions which gauge and reinforce their understanding of the material as it is presented. In an instructional context, these can provide the questions for computer lab activities that parallel the book. Lastly, the chapters conclude with an unguided task. This challenges the readers to reinforce and apply the concepts and skills from the introduction and the guided tasks to a new situation. These unguided tasks also have a dedicated debugging task included within most chapters following the chapter on debugging (Chap. 6) in order to develop and reinforce debugging skills. I encourage the reader to follow along with the guided tasks as you proceed through the chapter and take the time to debug the debugging exercises. These are based upon the most common bugs students have presented to me over the course of 13 years of teaching this material, so learning to recognize and fix these bugs can be very helpful to your success as a programmer. Lastly, taking the time to complete the unguided tasks will help reinforce the application of the principles within the chapter.

Together, the overall structure of the book, as well as the chapter structure mentioned above, seeks to prepare the readers for a long-term ability to do GIS programming, whether in industry or academic research. In particular, the intent of integrating computer science principles and the exercises into the same chapter is that a firmer understanding of the principles will help the reader transition from one language, such as VBA in ArcMap, to a new platform, such as Python in ArcGIS Pro. This understanding prepares readers for working in a dynamic, rapidly changing technology field.

Credits are due to Esri, both in general for the development of the software and in specific for use of the screen captures. Screenshot images are the intellectual property of Esri and is used herein with permission. Copyright © 2021 Esri and its licensors. All rights reserved.

As the dedication says, no project that even approaches the magnitude of a book happens alone. I would be remiss if I did not acknowledge those who have assisted me along the way to this guide, whether intentionally or unwittingly. Those whose influence is most direct are the 100-plus students who took the GIS programming courses I have taught from 2009 until the publication date of this book, especially those from 2019 to 2021, whose course text consisted of early drafts of Chaps. 1–17 of this book. They provided valuable feedback in the revision process. Those whose influence is less direct are the colleagues and especially the instructors I have had as an undergraduate and graduate student at Gustavus Adolphus College and Pennsylvania State University. Their modes of teaching both geography and computer science have undoubtedly filtered into how I teach, and thereby the content of this book. Especial gratitude for effective teaching in this regard goes to Max Hailperin and Mark Gahegan. Lastly, the most important influence comes from my family. They provided an environment and setting that allowed me to complete this book, whether through stepping up to help out (thank you, Jason Nolan!), general encouragement (thank you, Jocelyn and Evan, my stepchildren!), or the many ways my wife, herself a librarian, has guided me through the writing and publication process. This truly could not have come to completion without her.

Morgantown, WV, USA   Jamison Conley

Contents

Part I

Getting Acquainted with Python

This first section of the book introduces you to the basics and fundamentals of programming. While the ultimate focus of the book is through the lens of programming within a GIS environment, and Python in ArcGIS Pro more specifically, this section does not deal much directly with the GIS aspect of the book. Just as you need a basic understanding and foundation of concepts like how raster divides the world into a grid, and vector represents the world through the geometric primitives of points, lines, and polygons before moving into broader and more in-depth concepts of geographic information science, a basic understanding of the foundation of how computer programs are built is needed before moving on to the more in-depth concepts of scripting specifically for a GIS environment.

With this in mind, the first chapter introduces the processes of scripting and programming. It sets out the premise of why scripting is useful, even for those who may not develop new algorithms. Following that, the first chapter introduces the way of "thinking like a computer," because when you get down to it, scripting is translating the tasks you want to accomplish into a language that a computer understands. The next three chapters function as a group, and collectively introduce the building blocks of programming languages, starting with variables and operators, moving through control statements, and concluding with functions and objects. The remaining two chapters of this section develop skills that are beneficial to anyone who writes scripts and programs: being able to read and decipher scripts that others have written, which may contain commands you are unfamiliar with, and being able to interpret error messages and debug them to fix the script.

Introduction to Scripting and Programming

Introductory Comments: Scripting Versus Programming

Geographic information systems (GISs) are incredibly powerful tools, performing a wide range of functions for geographers. As Ron Abler and others wrote, "GISs are simultaneously the telescope, the microscope, the computer, and the Xerox machine of regional analysis and synthesis of spatial data" (1988). They allow us to see the big picture of complex global systems, as a telescope would. They allow us to delve into detailed analyses and descriptions, as a microscope would. It allows the copying and sharing of geographic data, like a Xerox machine. But first and foremost, GIS is the computer of spatial analysis. While the principles of spatial analysis and geoprocessing are not strictly dependent upon computers, and in some cases, predate computers, the GIS is the primary computational tool within geography. After all, with enough time and patience, any GIS operation could potentially be done with a sufficiently detailed gazetteer or almanac to look up data, a pencil to do any calculations, and a map to draw to communicate the results. Even so, some operations are absurdly impractical to carry out in these circumstances, and it is all much more feasible with a computational database in place of the gazetteer, a processor instead of a pencil, and a monitor or printer for the map.

Recognizing the central nature of computational technology and computer science to GIS, this course will introduce concepts and principles of computer science underlying geographic information science. These are presented through the lens and the skill set of scripting within a geographic information system, specifically Python in ArcGIS Pro.

Throughout the book, you will see two terms: scripting and programming. They are similar to each other, and if you ask a dozen programmers what the exact distinction is, you'll probably get at least a dozen different answers, some of whom will question whether there is a distinction anymore. Within this book, since there are different definitions of this distinction, the difference is in how the code is run, which is directly related to the process that you, the programmer, use instead of the specifics of the code itself. Scripting is the process of entering a small segment of code, often a line or two at a time, and having that code run immediately to see what the outcome is. As one reviewer of this book's proposal succinctly put it, "most people call it scripting when they are essentially gluing together code segments from Stackoverflow and/or Esri documentation."

However, there are limits to this approach to coding, as you move from simple tasks to more complex tasks that can rarely be done effectively in short code snippets. This means that you may want to have more structured and/or formal approach to constructing and executing your Python code. This structure is represented by algorithms, which are the focus of task 2 in this chapter, and forms the basis of the transition from scripting to programming. Programming is a more structured, or methodical, approach to implementing processes in computer code, in which you construct an algorithm, and then turn the conceptual algorithm into Python code.

It is important to note that this distinction, as fuzzy as it is, relates to how you, as the coder, approach the Python code you write, not the code that is produced. Python can be used for both scripting and programming. The code that you produce within a scripting project, "gluing together code segments" in the reviewer's words, can even be similar to the code you develop through programming strategies. Within this book, the distinction is most prominent later in the book, as programming skills form the basis of constructing new tools and extensions to ArcGIS Pro, which is the basis of Parts III and IV of this book. There are times when I will also use the term "coding," which can refer to either scripting or programming.

J. Conley, *A Geographer's Guide to Computing Fundamentals*, Springer Textbooks in Earth Sciences, Geography and Environment, https://doi.org/10.1007/978-3-031-08498-0_1

Task 1: Why Use Scripting?

If you are reading this book, you are probably already familiar with ArcGIS. As such, you are aware of both the power of GIS in its graphical user interface (GUI) and its limitations. For the power, think of the Toolbox in ArcGIS Pro. There are hundreds, if not thousands, of individual tools within the Toolbox. These can, in turn, be combined into workflows of several tools to create processes more useful and more powerful than any individual tool alone. To make the most of these tools and workflows, though, we might want to use scripting.

Imagine a scenario within a city planning department, answering a few questions about each parcel within applications to build a new structure. You would want to know (1) how much of the parcel is going to be covered by the new building, (2) how close the building is to the property line, and (3) for commercial structures only, if the submitted plan also has sufficient parking space for the building. The first question would require calculating two areas, perhaps through creating attributes and performing "Calculate Geometry" operations in the attribute table, and then finding the ratio of those two areas. The second would require finding distances, perhaps through the Near tool. The third would again involve calculating areas, but this time between the building size and the parking lot size. While each of these operations alone is not difficult, a large enough city would have enough applications that it would be tedious to carry it out between the attribute table and the toolbox over and over again for each application. A simple script that can take in the application with the polygon information, compute what is needed, and answer those three questions without any additional steps from the user would make the job easier.

Question 1: Describe another scenario with a workflow of at least three steps that would be frequently repeated, and therefore would benefit from scripting. Also provide a step-by-step set of instructions for the workflow.

The other limitation that can inspire scripting is that, as extensive as the toolboxes are, they cannot cover everything. New statistical techniques are developed on a constant basis, and you may want to use a statistical test which is not implemented in ArcGIS Pro. Scripting can also solve this, by allowing you to create a new tool, and write the code to carry out this new test. While the statistical toolboxes are growing within ArcGIS, they (at the time of writing) do not implement spatial lag or spatial error regression from the field of spatial econometrics (Ward & Gleditsch, 2008). If a person wants to use these tools, she or he would have to write a script to implement it in a new tool.

Question 2: Describe another tool you wish you had in ArcGIS Pro.

Task 2: Algorithms and Computational Thinking

The step-by-step set of instructions you provided in question 1 above is an example of an algorithm. Algorithms are the basis of how computer software works. They are sets of instructions that tell the computer precisely what to do. You can think of an algorithm like a recipe. A recipe for chocolate chip cookies tells you what you need to have—the flour, the sugar, the eggs, the chocolate chips, etc. It also tells you how much to use of each ingredient and what to do, and it provides a detailed set of instructions on how to combine and work with the ingredients to turn them into cookies. In this, the recipe is an algorithm for turning the flour, sugar, and other inputs into the output of cookies. Likewise, the city planning example I gave above would take as inputs the polygons for the lot, the building, and the parking area for commercial applications, and provide as output the answers to those three questions.

Most computers would not have an idea what your step-by-step instructions are above, however. You need to translate the instructions from English to "Computerese," which is the language that the computer can truly understand. Every programming and scripting language, from BASIC to Java to Python, can be considered a dialect of Computerese. A fuller description of Computerese, and Python, is in the following chapters, so only a few general ideas are introduced here.

First, the computer has zero common sense and a limited vocabulary. As an example, I'll draw from a new form of computer—the home voice assistant, like the Amazon Echo. Here is a conversation I had:

Me:	"Alexa, turn the air conditioning to 72 degrees."
Alexa:	"There is no device called air conditioning."
Me:	"Alexa, set the AC to 72 degrees."
Alexa:	"There is no device called AC."
Me [getting a little annoyed]:	"Alexa, set the temperature to 72 degrees."
Alexa:	"The air conditioning is set to 72 degrees."

Displaying the limited vocabulary, even though the phrase "air conditioning" appeared in Alexa's words to me, it did not have any meaning to the computer, as when I asked to set the air conditioning, it could not find a device called that. It only responded to "temperature." The computer also displayed a lack of common sense in not recognizing synonyms that any person would.

The second general principle of Computerese is that computers take everything literally. In the city planning example, I might have a typo in the ratio of the building size to the lot size, multiplying instead of dividing a building footprint of 3000 square feet by a parcel size of 9000 square feet to conclude that the building covers 27,000,000% of the lot. Any person will recognize that this is a nonsense answer, although the computer will happily report it back to the user. Even if you meant to divide, and multiplying gives a nonsense answer, the computer will carry out your instructions literally, and multiply the two numbers.

The third general principle, which is used under the hood, so to speak, of GIS, even if we don't see the practical effects often, is that the computer cannot see shapes or polygons. It can only work with numbers, and really only work with 0s and 1s. This means that everything in a GIS, from the text fields in an attribute table to all the geometries, must ultimately be converted into numbers. This has a bigger impact in, say, why it can take so long to find the intersection of two multipolygons, although it will not substantially impact the scripts we use and create in this course.

The fourth principle, which is the advantage, is that the computer can read and carry out these instructions incredibly fast. This benefit makes all the frustrations imposed by the first three principles worthwhile.

To take these principles and turn out English thoughts into Computerese instructions, we have to engage in computational thinking, trying to think like a computer might. A good quote on this comes from an unusual source: *Dirk Gently's Holistic Detective Agency* by Douglas Adams (1987):

> Richard continued, "What I mean is that if you really want to understand something, the best way is to try and explain it to someone else. That forces you to sort it out in your mind. And the more slow and dim-witted your pupil, the more you have to break things down into more and more simple ideas. And that's really the essence of programming. By the time you've sorted out a complicated idea into little steps that even a stupid machine can deal with, you've learned something about it yourself."

To program the computer, or give it the right instructions, we have to break the instructions down into very detailed, very simple ideas. We have to get to the building blocks of whatever language we are using, Python in this book. We can't just say find the area of a polygon; we have to tell the computer to access the geometry of the polygon, and from that geometry, get the area. We can't just tell the computer to calculate a ratio; we have to explicitly divide one number by the other. This means that an essential step in any programming task is to break a general task into smaller subtasks, and break each of those down into individual steps, and then translate each individual step from English into Computerese. This "divide and conquer" approach is a critical component of programming and computational thinking.

This is illustrated within the following scenario. Imagine you are a GIS consultant tasked with planning the stops for a series of rallies for a presidential candidate in the last week of the campaign. Your task is to find the sites of seven rallies, one per evening. The campaign manager has provided you with a set of constraints: (1) the rallies must be in different states, (2) the rallies need to be in states where the polling indicates a close contest, (3) the rallies need to be in large cities in those states, and (4) each rally needs a venue with at least 10,000 seats. Additionally, the campaign has provided you with the polling data, the city population data, and the available venues in each city, so that you have all the data you need to carry out the task.

At first, this might seem daunting, but through the "divide and conquer" strategy, it can be answered. Looking at the criteria, it would seem that there are two parts to this task, so we could have two subtasks. The first is to identify the seven states. Then, once we have the states picked, we need to identify a city and venue within each of the chosen states. Using this, we can start a very general outline.

A. Identify states for rallies.
B. For each selected state, choose a city and venue.

We have one main criterion for the states: that the polling is close. We have polling data, so we can find the difference in voting intent between the two top candidates. This might or might not be in the data itself, so we may have to calculate the difference as the absolute value of candidate A minus candidate B. Once we have this, find the seven states with the lowest values, which can be executed by sorting the states along this new variable and finding the lowest ones. This illustrates taking the subtask and breaking it down further into individual steps, as in the more specific outline below.

A. Identify states for rallies.
   a. Calculate new field in polling data:
      i. Margin = |Candidate_A – Candidate_B|/(Candidate_A + Candidate_B)
   b. Sort states by Margin (with the lowest values first).
   c. Select the first seven states in the sorted list.
B. For each selected state, choose a city and venue.

Now we can break the second subtask down. Within each state, we would need to find all the venues in the list of available venues which have a large enough capacity. This can give a list of venues with associated cities, and we would then want to identify the associated city with the largest population, and pick the biggest venue in that city. You will notice a process of initializing variables, or setting them to a starting value, and updating them as I cycle through a list of venues. The initialization process is what happens in steps B.b and B.c. The updating portion happens in steps B.d.i.1 and B.d.i.2. This is a common programming strategy as well. Filling this out within the outline gives what is below.

A. Identify states for rallies.
   a. Calculate new field in polling data:
      i. Margin = |Candidate_A – Candidate_B|/(Candidate_A + Candidate_B)
   b. Sort states by Margin (with the lowest values first).
   c. Select the first seven states in the sorted list.
B. For each selected state, choose a city and venue.
   a. Select venues within the state from the list of all venues in the country.
   b. Set the rally venue to the first venue in the list.
   c. Set the rally city to the city containing the rally venue.
   d. For each venue within the state:
      i. If it has a capacity above 10,000 seats:
         1. If its city's population is more than that of the current rally city, change the rally city and rally venue to this venue and its city.
         2. If its city's population is equal to that of the selected city (which probably means the same city), change the rally city and rally venue to this city and venue **only if** the venue's capacity is larger than the capacity of the current rally venue.

This outline helps to organize the process of translating English tasks into Computerese, and illustrates how the overarching task is broken down into simpler and simpler steps. The outline is in what can be called "pseudocode," which is a generic form of Computerese, rather than any specific scripting or programming language. It gives the set of instructions that will be carried out within the algorithm to complete the task, and these pseudocode instructions can be transcribed into any scripting or programming language. Having this intermediate step of detailed instructions is useful, and is therefore something that I would recommend you continue to do in the unguided exercises throughout this book.

Task 3: Non-guided Work

As the unguided task, imagine yourself doing GIS work for a conservation agency, seeking to propose areas to set aside for the protection of an endangered bird species. You have the following data: known nesting sites of the bird, a polygon-based dataset of forest areas, and a dataset of oil and gas wells. The proposed areas are forest areas which are at least 5 square miles in area, contain nesting sites, and which have no oil and gas wells either inside them or within 0.5 miles of the edge of the forest. Your task is to identify all forests meeting these criteria.

Question 3: Provide a detailed outline of instructions for carrying out this task. Use at least two levels of the outline (outline the task into subtasks, with individual steps inside each subtask). You do not need to turn the individual steps into Python (or any other language).

Use the example outline for locating political rallies at the end of Task 2 as a guide for the level of detail that is expected.

Basics of Programming: Variables, Operators, and Statements

2

Introductory Comments: Parsing a Line of Code

Programming in Python is much like writing in any other language, and this chapter begins the development an analogy of writing in Python to writing in a foreign language, which, in a way, is what scripting and programming are. That language is the Python dialect of the language introduced in Chap. 1 as Computerese. All programming languages are dialects of this language of Computerese, because they are all that similar to each other. While this book uses Python, the concepts are applicable to all programming and scripting languages. Chapter 19, Task 5, illustrates this when attaching a script written in a different language to an ArcGIS Pro tool. Therefore, if/when ArcGIS no longer uses Python, but switches to a different language, the skills you develop in this exercise will translate and transfer to whatever comes next.

To focus on the basics of Python, this chapter, like the remainder of this section, will not use any ArcPy functions. The scenario is managing the roster for a GIS training seminar.

This chapter begins by looking at how a computer reads a single line of code, equivalent to a phrase or sentence of Computerese.

Every time a computer encounters a line of code to run, it first has to figure out what those instructions are. After all, as said in Chap. 1, everything is broken down into 1s and 0s, including text, meaning even the program itself must be transformed into a series of 1s and 0s. We won't go anywhere near as far as the 1s and 0s, but we will look at the first step in that process of breaking down a line of computer code. As will be seen in Chap. 6, many of the bugs we can encounter are the result of the computer not being able to understand a line of computer code, so an understanding of how the computer approaches this process will help you recognize the nature of the bug and thereby fix it. A GIS analogy for this is recognizing that all the vector data in a GIS is, as far as the geometry is concerned, built upon the basics of points, lines, and polygons. Even though you might not think explicitly about the point coordinates when carrying out an analysis, having an understanding of how those points work and the properties of them, like what coordinate system and projection they are using, can help you to identify and troubleshoot different problems of analysis, such as when two datasets that should overlap don't because they are using different coordinate systems. Likewise, understanding how the computer breaks down the script into those building blocks, and the properties of those building blocks, can help you debug and troubleshoot error messages that arise when you are writing a script.

The first step in breaking down a line of code, or parsing it, is to identify the constituent parts of that line. This is equivalent to having to break a sentence down into the individual words that make up that sentence. Parsing is, in other words, how the computer reads Computerese. In computer science, these individual parts are called tokens, and different kinds of tokens fulfill different aspects of Computerese, much like nouns, verbs, and adjectives all have different roles within spoken language. I'll illustrate the premise of parsing and breaking down the code with a few examples below, drawn from the hands-on tasks which appear after these introductory comments. In those guided tasks, you will see how to enter these Python lines, and others, into ArcGIS Pro to gain experience with these ideas.

The first line is the following. Note that I will use the `courier new` font for Python code, which is commonplace for representing computer code in any language because it has the advantage that every character is the same width.

```
student_price = price * 0.5
```

© The Author(s), under exclusive license to Springer Nature Switzerland AG 2022
J. Conley, *A Geographer's Guide to Computing Fundamentals*, Springer Textbooks in Earth Sciences, Geography and Environment, https://doi.org/10.1007/978-3-031-08498-0_2

Each line, no matter how long or short it is, gets broken into these tokens. This is a simple line, which you can probably deduce from a little algebra sets the student price of something at half the regular price. The computer, though, lacks the common sense to figure it out, and must read it, token for token. However, first, the computer has to figure out what the individual tokens are in this line. This is a straightforward splitting of the line based upon where the spaces are. I've divided it into tokens using background shading. Each switch between light and dark background indicates a new token.

```
student_price = price * 0.5
```

Once this is done, the computer recognizes `student_price` and `price` as names. While we might recognize this by looking at it as them being variables, the computer doesn't know it yet. It does, though, automatically recognize = and * for the mathematical operations they represent, and `0.5` as a number. Based upon how the = operator works, the computer either creates a new variable called `student_price`, or it changes the value already assigned to that variable. Meanwhile, `price` must be a variable, and one that is a number at that. Otherwise, there will be an error, the nature of which will be covered in Chap. 6 on debugging.

The second line adds a new element, in which not all words are names.

```
paid = True
```

However, the idea of splitting the line into tokens at each space still remains. This gives the following breakdown:

```
paid = True
```

As with the first line, it creates or reassigns the value of a variable called `paid`. This time, though, the other word, `True`, is not a variable, but is itself what is called a keyword, or a word with a very specific meaning.

The next example extends what is a token by introducing the quote marks. These must be present in pairs, because when the computer encounters the first quote mark, it immediately finds the corresponding ending quote mark and treats everything inside, along with the quote marks themselves, as a single token representing a piece of text.

```
name = "ArcGIS Training"
```

The following breakdown has the quote marks underlined to indicate the pairing of the two together.

```
name = "ArcGIS Training"
```

The next line builds upon the idea of the quote marks, with a pair of parentheses. It also adds commas. You can probably recognize this as a list of names.[1]

```
participants = ("John Smith", "Fred Jackson", "Alice Scholl", "Bob Waters", "Eve Douglas", "Monty
Adams")
```

The parentheses, like the quote marks, function as a pair. If you have one without the other, there will be an error message. While the quote marks indicate that everything inside of them is a text string, the parentheses tell the computer that everything inside of them is a list. Each of the items in the list is separated by a comma. The commas and parentheses are deemed here as their own tokens, because they fulfill roles in the line of code that are independent of everything around them.

```
participants = ("John Smith", "Fred Jackson", "Alice Scholl",
"Bob Waters", "Eve Douglas", "Monty Adams")
```

---

[1] As will be clear later, it is not, in strict Python terminology, a list. At this point, though, we can colloquially call it a list, so throughout these introductory comments, I'll refer to it as a list.

Also, having the more complex line begins to illustrate some more principles of how the computer parses a line, particularly the way that it depends upon patterns in the sequence of tokens to figure out what the line is doing. In this example, there are a few elements in the syntax, or structure, of the line to tell the computer that this is a list. The first of these is the matched set of parentheses, which here indicate the start and finish of the list. The other part is that what is in between those parentheses is a set of things—text strings in this case—each separated by a comma. It also determines what is on the right side of the equals sign before it attempts to place that value in the variable on the left side of the equals sign.

The next example shows that the parentheses can also impact how other tokens get interpreted.

```
print(student_price)
```

The parentheses and its contents function much the same way as above, although in this case, it is only a single item in the list.[2] However, unlike the previous line, which took the list of names and set it to be the contents of a variable, this one has no = operator to indicate that.

$$print(student\_price)$$

Because there is no operator, this changes the syntax of the line, which in turn changes how the computer interprets the word print. Because the word print is followed immediately by the opening parenthesis, the computer recognizes print as a function, not a variable. We will see more detail on functions in Chap. 4.

The last example here extends this with one more token: a dot.

```
participants.reverse()
```

In this case, the dot, however, doesn't represent a decimal point, but separates two words. Also notice the empty list at the end of the line, as represented by the two parentheses with nothing between them. Both the dot and the parentheses are critical to the correct running of this line of code.

$$participants.reverse()$$

The dot signifies that the token before the dot, participants in this example, must be a variable, and not just any variable. It has to be of one of a set of particular types of variables, called objects. What, then, is after the dot is something similar to a function, which is what the parentheses indicate. While both functions and objects will be used starting in this chapter, Chap. 4 will have much more information on both functions and objects.

Task 1: Basics of Python

Variables, Literals, and Operators

Before we begin our examination of Python in ArcGIS Pro, we need to know how to access Python. There are three primary ways ArcGIS Pro interacts with the Python programming language: the Python Window, Python Notebooks, and IDLE. We will start with the Python Window, as it is the simplest means of interacting with small pieces of Python code. In other software environments, you may have encountered the term "Python Shell," which is what this essentially is. It allows you to enter a statement in Python, and that statement will be executed immediately. This can be useful for exploring what different elements within the language do, so we will be using it for pedagogical purposes here. The next level of complexity is the Python Notebook, also integrated with recent versions of ArcGIS Pro. It allows you to work with snippets of code which can be multiple statements, and execute them in batches. The practice of Python scripting is moving toward the notebook para-

---

[2]Just like you may well go to the grocery store with a list of one item: milk, a list in Python can have only one item. It can even have zero items, which is useful if we want to build a grocery list from scratch, one item at a time. We start with an empty grocery list, or a blank piece of paper, and then write items on it, like milk, eggs, and bread.

digm, thanks in part to a scripting environment called Jupyter,[3] which is the underlying technology for the Python Notebook in ArcGIS Pro. We will use it intermittently with the Python Window starting in Chap. 3. The third approach for ArcGIS Pro here is IDLE, which stands for "Integrated Development and Learning Environment" and is a programming environment that is included in the installation of ArcGIS Pro. It supports the most complexity in programming and debugging, and as such, when we shift from scripting to programming to support writing new Python-based tools to enhance ArcGIS Pro, we will use IDLE more extensively. To show its potential for investigating and debugging code, IDLE is introduced in Chaps. 5 and 6.

Recall from the introductory comments that the scenario here is managing the roster for a GIS training seminar. The first thing we may want to do is set up a few basic properties of our seminar, like the name, address, and registration cap. In Computerese, these are equivalent to nouns. These can all be *variables* within our script. Variables are containers that hold values, and are referred to by names. You can think of a variable as a box that has a label on it for its name (Fig. 2.1).

Start ArcGIS Pro with a new project. Select "Map" under a new "Blank Template." Give it a sensible name, like Exercise2. Go to the Analysis ribbon and open the Python Window. If you have a version of ArcGIS Pro that is sufficiently updated to have both the Python Window and Python Notebook, make sure you are using the Window.

First, a note about the Python Window you now see. It has two parts. The lower part, which is currently one line, reads "Enter Python code here." Here, you will enter Python code line by line, and it gets executed as soon as the statement is completed. Chapter 3 shows how statements can be longer than a single line. The results of executing the statements you enter below are given in the top part of the Window.

Creating variables in Python is straightforward. All we need to do is type the following line:

- `name = "ArcGIS Training"`
- Hit the Enter key.

Now the line appeared above, and the code entry line is once again blank. This means that the previous line was executed.

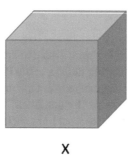

Fig. 2.1 The x variable

---

[3] See http://jupyter.org for more information.

You don't see anything else, because the line simply created the variable, and gave it a value. It did nothing with the variable. We now have a box labeled *name* with the content "ArcGIS Training" (Fig. 2.2).

If we want to do something with it, we can print the value of the variable out with a `print` function.

- In the entry line, type `print(name)` and press Enter.
  It printed out the value of the variable name, which we just set to "ArcGIS Training."

Instead of constantly repeating the image of the Python Window from here on out, I will represent the output with the executed statement first following three > characters, and then the output (if any). Those who have used other programming shells, including the one in IDLE, may recognize this convention.

```
>>> print(name)
ArcGIS Training
```

Without getting too far ahead in the analogy of Computerese, the lines you just entered were both statements, or sentences, within Python. As with English and other languages, and as stated in the introductory comments, all statements have a specific grammar, or syntax, that they must adhere to.[4] The syntax of setting up a variable is as follows, with at least three

Fig. 2.2 The `name` variable

---

[4]Or "to which they must adhere," depending upon the grammar rules you prefer. As we will see later, especially in Chap. 6, humans reading English can handle flexibility and ambiguity like this. Computers reading Python cannot.

tokens, the first of which is the name of the variable, the second of which is the equals sign, and the remainder of the line is an expression which will evaluate to whatever you want to store in that variable.

```
[variable name] = [value]
```

In this syntax, the equals sign is an *operator*. Operators act in many ways as the verbs in Computerese. Just as a sentence without a verb does nothing (and is typically not permitted), operators hold the actions that the computer will take. In this case, the operation is setting a value for a variable. Other common operators are mathematical operations (like +, −, etc.).

- Use this syntax to create two more variables corresponding to the following boxes.
  - Note that for numbers, you do <u>not</u> need quote marks. In fact, the computer will read "9" as a text string containing the character 9 instead of a number (Fig. 2.3).
  When complete, you should have the following:

```
>>> address = "98 Beechurst Ave."
>>> cap = 20
```

You may have noticed that we have two different kinds of variables here: two text strings and a number. In some programming languages, you have to specify the *type* of variable, such as text string, integer, or real number. Python handles this for you, so you do not have to specify, although the types do nonetheless exist, and can limit what is or isn't possible within a line of code. To illustrate this, let's create one more variable: price. At first glance, for printing brochures, you might want the price to be a string that includes the dollar sign.

- Type `price = "$100.00"`
- Hit enter.

After some consideration, you realize you might have to calculate discounts, such as a 50% student discount. Let's try multiplying the price variable by 0.5.

- Type `student_price = price * 0.5` and hit Enter.
  You should have an error message, undoubtedly the first of many that you will encounter as you learn programming.
Question 1: Use your own words to paraphrase the last line of the error message, about why the error happened. (In other words, do not simply transcribe the error message.)

```
>>> price = "$100.00"
>>> student_price = price * 0.5
Traceback (most recent call last):
  File "<string>", line 1, in <module>
TypeError: can't multiply sequence by non-int of type 'float'
```

Fig. 2.3 Two more variables

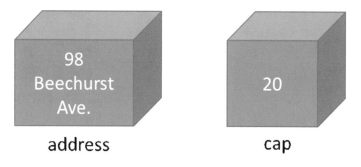

address   cap

Python handles the bookkeeping of which data type is contained in each variable behind the scenes, making it very easy to change the type of the variable. We simply reassign the value. This is equivalent to taking the text string "$100.00" out of the box labeled price and putting the number 100 into it.

- Type `price = 100.00` and press Enter.
- Now we want to reenter `student_price = price * 0.5`. You can press the up arrow key twice to bring back this line, as the up and down arrow keys navigate forward and backward in the history of the Python statements you have entered during this session. Once you have it, press Enter.
- Lastly, type `print(student_price)` and press Enter to confirm its value.

```
>>> price = 100
>>> student_price = price * 0.5
>>> print(student_price)
50.0
```

There are several data types in Python, and all programming languages, for that matter. Some of the most common ones are text strings; integers, which in Python are called short and long types;[5] real numbers, which in Python are float and double types; Boolean variables, which are True or False; sequences, which will be covered in more depth shortly; the `none` type, which corresponds to an empty box; and other, more complex custom-designed object types, such as an ArcGIS feature class or map legend. Continuing the box analogy, we can represent them with different colored boxes. Text strings are green; numbers are blue, with integers darker than real numbers; a Boolean value is orange; and the `none` type is red (Fig. 2.4).

Fig. 2.4 A set of variables, color-coded by type

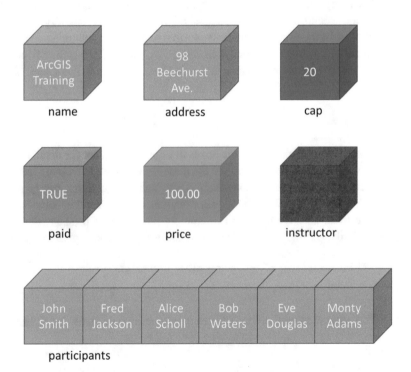

[5]There are different integer types with obscure names, as well as different real number types with obscure names for the following reason. All numbers have to be represented in binary—that is, 0s and 1s. Computers are designed such that each number should be a consistent number of binary digits (aka bits) long. In the early days of computer programming, they figured that 32 bits would be sufficient, as it can take 2^{32} distinct values, ranging from about -2.15 billion to 2.15 billion. Engineers and programmers soon realized that this wasn't enough. Because many operations were developed with a 32-bit number in mind, they created a different data type that is 64 bits. To distinguish them, they are called short integers and long integers, which gives you the data types *short* and *long*. Real numbers are more complicated. They break the 32 bits into two parts and use scientific notation, representing 504.16 as $5.0416 * 10^2$. The larger part is the mantissa (5.0416) and the smaller part is the exponent (2). Because of the exponent, the decimal point can be considered to "float" within the number, rather than be a fixed location. Hence, it is called the *float* type. As with the integers, it became apparent that 32 bits was insufficient, and a 64-bit real number type would be needed. Since it has double the number of bits as a float, it is called the *double* type.

Text strings are in green; numbers are in blue, with the real number in lighter blue; and the integer in darker blue. The Boolean value `paid` is either `True` or `False`, and is in orange. The `instructor` has not yet been determined, so is an empty box, representing the `None` type. Lastly, we have a sequence of `participants`, which is a series of names, or text strings, all joined together in a sequence of boxes.

To create the variables for paid and instructor, enter the following lines. Note that `True` and `None` <u>are</u> case-sensitive keywords, as is the other Boolean value, `False`.

- `paid = True`
- `instructor = None`

Before going any farther, here is a note on variable names. They can contain letters, numbers, and underscores, but cannot start with a number. Also, they cannot be *keywords*. Keywords are words that have been reserved for a very specific purpose within Python. You now have two keywords on the screen, and they are in slightly darker font—**True** and **None**. Because they are reserved for specific values in variables, you cannot have variables called True or None. We will see keywords throughout this book, and a complete list is available at https://docs.python.org/3/reference/lexical_analysis.html#keywords.

There's one further note for assigning variable values. We have, thus far, been using another form of token, *literals*, which are values that the computer will interpret exactly as they are. In that sense, literals are another form of nouns for Computerese. Examples are numbers (100, 0.5, etc.) and text strings ("ArcGIS Training," "98 Beechurst Ave.," etc.).

You might, at times, want to do something like the following, in which a variable exists on both sides of the equals sign:

- `x = y`

The computer can handle this in two different ways, and some programming languages choose one approach, while others chose the second. The first approach is to have both names, `x` and `y`, refer to the same box. This is not what Python usually does. Most of the time, Python instead still treats them as separate boxes, but copies the *contents* of `y` and puts that value in `x`. Therefore, if you do this, and then change the value of `x`, `y` will still have that original value, instead of change along with `x`. However, there are exceptions to this in Chap. 12, Task 3, and Chap. 20, Task 5.

Sequences

We now have variables for all the boxes in Figure 2.4 except for the list of participants, which is a kind of sequence. Python has several kinds of sequences, and the three we will look at here are lists, tuples, and dictionaries. Lists and tuples are both ordered, meaning the items they contain have an index, and thus can be referred to as the first, second, third, etc. item in the list. The only difference is that tuples cannot be edited after they have been created, while lists can be changed. This difference restricts what methods are available for tuples versus lists, as things which would alter the contents as an effect of the method, like sorting a list, are not available for tuples. The syntax for creating a list and a tuple is as follows:

```
list_name = [item_0, item_1, item_2, …, last_item]
tuple_name = (item_0, item_1, item_2, …, last_item)
```

You will note that the only difference is square brackets for the list and parentheses for the tuple. Let's create a tuple of `participants` with the following code (all on one line):

- `participants = ("John Smith", "Fred Jackson", "Alice Scholl", "Bob Waters", "Eve Douglas", "Monty Adams")`
- Press Enter.

Tuples are ordered, so we can use an index to access individual items in the tuple. Perhaps counterintuitively, the indexes start at 0, rather than 1. Thus, John Smith is the zeroth item in the tuple. To access items in the tuple, you enter the name of the tuple, followed by the index you want in square brackets. (This syntax uses square brackets for both lists and tuples. If you use parentheses, you will get what is, at least for now, a very confusing error message. If you recall the syntax patterns described in the introductory comments, remember that a name that immediately preceded parentheses was interpreted as a function, rather than a variable. If you use parentheses here, it will try to interpret the name of your sequence variable as the name of a function, when no such function exists, and the computer gets very confused by this.)

- Type `participants[0]` and press Enter.

```
>>> participants = ("John Smith", "Fred Jackson", "Alice Scholl", "Bob Waters", "Eve Douglas",
"Monty Adams")
>>> participants[0]
'John Smith'
```

You will see "John Smith" printed out. It did this even though we didn't use the print function, because if you type into the Python Window an expression that evaluates to a specific value, instead of a statement that doesn't, the Python Window will automatically print out the value of that expression. The distinction between expressions and statements is discussed more fully in Task 1.5 of this chapter.

Accessing a subset of the tuple is through a more complex command. We replace the single index, which is 0 above, with a range. The range is denoted as follows, introducing a new character, the colon: `start_index:finish_index_plus_one`. Alice is in index #2, and Bob is in index #3. To access the subset of Alice and Bob, we want the range 2:4, because it is the last item we want plus one (or, put another way, the index of the first item <u>not</u> in the subset).

- Type `participants[2:4]` and press Enter.
  You will see that it returned a tuple with two items: Alice and Bob.

```
>>> participants[2:4]
('Alice Scholl', 'Bob Waters')
```

If we want to access the last item in the tuple, but do not know how long the tuple is, we can use a negative index. The index -1 refers to the last item in the tuple, -2 refers to the next to last item, -3 is the one before that, and so on.

- Type `participants[-1]` and press Enter.
  You will see that this returned Monty Adams.

```
>>> participants[-1]
'Monty Adams'
```

We may want to find out how many people have registered. There are functions for sequences that we can use to find this, among other things.

- Type `len(participants)` and press Enter.
  It returns 6, which is the number of items in the tuple.

```
>>> len(participants)
6
```

To continue the scenario of managing the participants list for a GIS training, imagine you get an email from Eve asking if her registration was included. Using the `in` keyword, we can answer this question.

- Type `"Eve Douglas" in participants` and press Enter.
  It returns `True`, because she has registered.

```
>>> "Eve Douglas" in participants
True
```

You then get another email from Tyler Rose, asking if he has registered.

- Use the same syntax to confirm that Tyler is not in the tuple of participants.
  Since he has not registered, we want to add him to the tuple. Adding items to sequences is through the `append` command.[6]
This command is specifically applied to the variable it is operating on, so its syntax includes both the sequence name and the
item to be appended. Its syntax is as follows. Both the period and the parentheses are required.

```
name.append(item)
```

- Type `participants.append("Tyler Rose")` and press Enter.
  You will get an error message saying that tuples do not have the attribute of append.

```
>>> participants.append("Tyler Rose")
Traceback (most recent call last):
 File "<string>", line 1, in <module>
AttributeError: 'tuple' object has no attribute 'append'
```

Question 2: Why do tuples not support the append method?
Therefore, we need to change `participants` from a tuple to a list.

- Type `participants = ["John Smith", "Fred Jackson", "Alice Scholl", "Bob Waters",
  "Eve Douglas", "Monty Adams"]`.
- Press Enter.
- Now reenter the line `participants.append("Tyler Rose")` and press Enter.

  You will notice that there was no error message.

- To confirm Tyler was added, type in `participants` and press Enter.

```
>>> participants = ["John Smith", "Fred Jackson", "Alice Scholl", "Bob Waters", "Eve Douglas",
"Monty Adams"]
>>> participants.append("Tyler Rose")
>>> participants
['John Smith', 'Fred Jackson', 'Alice Scholl', 'Bob Waters', 'Eve Douglas', 'Monty Adams', 'Tyler
Rose']
```

Next, you find out that Bob Waters would prefer to be called Robert. We can change the item in the list. First, we might
want to know where in the list Bob's name is located. This is through the `index` method.

- Type `participants.index("Bob Waters")`.
- Press Enter.

  It printed out the index for you.

```
>>> participants.index("Bob Waters")
3
```

---

[6]I use the term "command" here deliberately. It is neither an operator nor a function, but is technically a method attached to the variable called
name, which is an object of type "list." Objects will be treated with far more detail in Chap. 4.

Let's try to create a new variable to make this change. Type the following lines:

```
>>> bob = participants[3]
>>> bob = "Robert Waters"
```

Type `participants` and Enter to print the list.

```
>>> participants
['John Smith', 'Fred Jackson', 'Alice Scholl', 'Bob Waters', 'Eve Douglas', 'Monty Adams', 'Tyler Rose']
```

Question 3: Why did this not change the list?
Now to change the item, we have to edit directly in the list.

- Type `participants[3] = "Robert Waters"` and press Enter.
- Type `participants` and press Enter to print the list and confirm the change.

```
>>> participants[3] = "Robert Waters"
>>> participants
['John Smith', 'Fred Jackson', 'Alice Scholl', 'Robert Waters', 'Eve Douglas', 'Monty Adams', 'Tyler Rose']
```

As we construct our roster of participants, we may want to place them into alphabetical order. This can be done through the `sort` method.

- Type `participants.sort()` and press Enter.
- Type `participants` and press Enter to print the list and confirm the change.

```
>>> participants.sort()
>>> participants
['Alice Scholl', 'Eve Douglas', 'Fred Jackson', 'John Smith', 'Monty Adams', 'Robert Waters', 'Tyler Rose']
```

Sorting is always done from A to Z for strings, or from low to high for numbers. If you wanted to have it from Z to A, then you can use the `reverse` method.

- Type `participants.reverse()` and press Enter.
- Type `participants` and press Enter to print the list and confirm the change.

```
>>> participants.reverse()
>>> participants
['Tyler Rose', 'Robert Waters', 'Monty Adams', 'John Smith', 'Fred Jackson', 'Eve Douglas', 'Alice Scholl']
```

The last form of sequence that we will look at is a dictionary. Dictionaries are not ordered, but they have pairs of items associated with each other, much like a dictionary associates words with their definitions. It doesn't matter what the 42nd word in the dictionary is, as long as you can find its definition. In this scenario, we would want to associate each participant with whether or not they have paid yet.

Constructing a dictionary can take several steps.[7] We will first set up an empty dictionary. Where tuples use parentheses, and lists use square brackets, dictionaries use curly braces.

- Type `paid:dictionary = {}` and press Enter.

We now need to enter each person with their payment status. Adding an item to a dictionary is through the following syntax:

```
dict_name[key] = value
```

Note, again, the use of brackets instead of parentheses or curly braces. In this example, the key is the participant's name, and the value is their payment status. Enter the following lines, pressing Enter after each one:

- `paid:dictionary["Alice Scholl"] = True`
- `paid:dictionary["Eve Douglas"] = False`
- `paid:dictionary["Fred Jackson"] = False`
- `paid:dictionary["John Smith"] = True`
- `paid:dictionary["Monty Adams"] = True`
- `paid:dictionary["Robert Waters"] = False`
- `paid:dictionary["Tyler Rose"] = True`
- To look at the dictionary, enter `paid:dictionary` and press Enter.

```
>>> paid:dictionary = {}
>>> paid:dictionary["Alice Scholl"] = True
>>> paid:dictionary["Eve Douglas"] = False
>>> paid:dictionary["Fred Jackson"] = False
>>> paid:dictionary["John Smith"] = True
>>> paid:dictionary["Monty Adams"] = True
>>> paid:dictionary["Robert Waters"] = False
>>> paid:dictionary["Tyler Rose"] = True
>>> paid:dictionary
{'Alice Scholl': True, 'Eve Douglas': False, 'Fred Jackson': False, 'John Smith': True, 'Monty
Adams': True, 'Robert Waters': False, 'Tyler Rose': True}
```

You can see that there are pairs of keys and values, printed together with colons in each pair. While in this case, they were printed in the same order in which they were added, this is not guaranteed because dictionaries, unlike lists and tuples, do not have an inherent ordering.

To get a list of the keys in a dictionary, which in this case is a list of participants, a two-step process is needed. It starts with the `.keys()` method of the dictionary, but this does not return a list. It instead returns something called a View, which can be turned into a list through the `list()` function. Since we have no need of the View, we don't need to create a variable to hold it. Instead, we can take that and put it directly into the list function as below. This highlights another element of parsing more complex lines of code. The input to a function is what is between the parentheses following the function name of list. Just like in algebra, you evaluate things inside parentheses before doing anything else. That means the expression of `paid:dictionary.keys()` gets evaluated, and the result of that evaluation gets put into the `list()` function.

- Type `list(paid:dictionary.keys())` and press Enter.

---

[7]This is one way of creating the dictionary. There are others that can create it in a single, lengthy, line of code. I prefer this approach, especially as it can work well with loops, which we will see in the next chapter.

```
>>> list(paid:dictionary.keys())
['Alice Scholl', 'Eve Douglas', 'Fred Jackson', 'John Smith', 'Monty Adams', 'Robert Waters',
'Tyler Rose']
```

To find out the value for a particular key, use the key in brackets, much like you would put the index in a list or tuple in brackets.

- Type `paid:dictionary["Tyler Rose"]` and press Enter to find out if he has paid yet.

```
>>> paid:dictionary["Tyler Rose"]
True
```

This section introduced three types of sequences: tuples, which are ordered and cannot be edited; lists, which are ordered and can be edited; and dictionaries, which have no order, but instead pair keys with values. There are other types of sequences within Python, although we will only sparingly encounter them in this book.

Strings

Just as there are a series of commands specific to lists, tuples, and dictionaries, there are commands specific to text strings. Let's start with the basics, working with the title of the training. Recall that we stored it in the variable called `name`.

- To remind yourself of the training course name, type `name` and press Enter.
  If we want to see whether a particular word, letter, or any set of characters is in the string, we can use the keyword `in`.

- Type `"GIS" in name` and press Enter.
  As you would expect, it returns True.

```
>>> name
'ArcGIS Training'
>>> "GIS" in name
True
```

If we want to know where in the string "GIS" is found, we can use the `find` method.

- Type `name.find("GIS")` and press Enter.
  It returns 3, which is the position in the string where the substring we were looking for ("GIS") <u>starts</u>. As with list indices, we start counting at 0 instead of 1.

```
>>> name.find("GIS")
3
```

Next, you are told that many people in the target audience for the training recognize the company name (ESRI) rather than the product name (ArcGIS). Therefore, you want to edit the name accordingly.
First, we can try to find "ESRI" in the name.

- Type `name.find("ESRI")` and press Enter.
  It returns −1, which is not telling you that ESRI starts at the last character of the string, but instead is a way of telling you that it wasn't found at all.

```
>>> name.find("ESRI")
-1
```

To change the name, we can use the replace function, which is straightforward.

- Type `name.replace("ArcGIS", "ESRI")` and press Enter.

```
>>> name.replace("ArcGIS", "ESRI")
'ESRI Training'
```

- Type `name` and press Enter.

```
>>> name
'ArcGIS Training'
```

You will see here that the replace function created a new string, but did not directly alter the old one, keeping the variable `name` as it was. The line we use for this shows a common practice in scripting and programming that might look unusual to the more mathematically minded. The variable used to calculate the expression on the right side of the equals sign is the same variable that is receiving the value of that expression on the left of the equals sign: `name = name.replace("ArcGIS", "ESRI")`. This works because, while you would read the line left to right, the assignment operator evaluates the right side first, and then takes whatever value that returns and places it in the box for the variable on the left. It will do this even if it means overwriting the value that was used in the expression.

- To save the change, type `name = name.replace("ArcGIS", "ESRI")` and press Enter.
- To confirm the save, type `name` and press Enter.

```
>>> name = name.replace("ArcGIS", "ESRI")
>>> name
'ESRI Training'
```

Lastly, seeing the company logo, you wonder if you need to keep things lowercase. This is done through the `lower` method, and uses the same form of assignment statement.

- Type `name = name.lower()` and press Enter.
- Type `name` and press Enter to see the change.

```
>>> name = name.lower()
>>> name
'esri training'
```

These are just a few of the commands for strings. I have presented some of the commonly used ones, and we will encounter additional commands later in the book.

Before continuing, I will introduce three terms which you may frequently encounter as you progress both in this book and in your online searches for sample code: delimiters, expressions, and statements. Building upon the analogy of Computerese as a foreign language, we can regard variables as our nouns, and operators as verbs, and keywords as special vocabulary words. Delimiters are the punctuation marks of Computerese. Expressions are like phrases, and statements like sentences. The next paragraphs will introduce these in more detail.

Delimiters

Delimiters are punctuation marks that guide the computer in interpreting the code you write. In this, they function like punctuation in English or other languages. However, imagine the computer as the worst stickler for grammar you have ever seen, worse than the high school English teacher who spent an entire week on when you should and should not use a comma. This is because humans have common sense that (most of the time) allows us to figure out what the author was trying to say, even if the author did not follow the Chicago style guide precisely. Returning to the computer as the high-speed idiot, it lacks that common sense. If a comma is missing, it truly cannot figure out what you are trying to tell it. Some example delimiters that we have seen before are parentheses for tuples, brackets for lists, curly braces for dictionaries, colons in index ranges (and we will see them more in complex statements, coming in the next chapter), commas separating items in a list or tuples, and periods noting methods specific to objects, such as the period in the `name.lower()` command you just typed in. A common source of errors is missing or incorrect delimiters, so it is worth paying attention to them. A full list is available at https://docs.python.org/3/reference/lexical_analysis.html#delimiters. That list also classifies assignment operators, like the = sign, as delimiters that "also perform an operation." For the purposes of this book, to be less confusing about this dual purpose, I have introduced them as operators. Likewise, quote marks are treated as special characters, but are not deemed to be delimiters because they are part of the token with the rest of the string that they contain, instead of a distinct token unto themselves.

Expressions and Statements

Expressions, as noted above, are analogous to phrases in standard writing. They evaluate to something, whether that is a number, a text string, a Boolean (true or false) value, or a more complex object. Some simple examples are:

```
x + 5
x == 5
name.lower()
```

Each of these evaluates to a single value, but does not provide a command for the computer to execute. This is the purpose of a statement, which is like a sentence. Just as it can be a challenge to come up with a concise definition of a sentence, the closest to a good, concise definition of statement is "a complete unit of execution." This definition is so standardized that it has, in fact, been circulated so many times online that finding the official source of this definition to cite is not possible. However, this definition is also vague enough that a lot can fall under this definition. Simple statements often, but not always, involve setting the value of a variable. A decent, but not foolproof, test you can use is whether it causes something to get printed in the Python shell you have open for the exercise. The shell will print the result of evaluating an expression, but it will simply execute a statement without printing anything (unless, of course, it is a print statement).

This chapter introduced you to the basic building blocks of Python, from the individual words that make up a Python statement, as variables, operators, and keywords. Different specific types of variables, including lists, tuples, dictionaries, and strings, were all introduced. Then this task wrapped up by introducing the concepts of delimiters, expressions, and statements. The next chapter builds upon statements by providing complex statements to give more nuanced control of what the computer executes.

Task 2: Finding and Fixing Bugs

One lesson that everyone working in any form of coding discovers extremely quickly is just how commonplace bugs are.[8] Therefore, being able to debug the code that you write is quite possibly the most important skill in any computing endeavor—more important than the ins and outs of any specific language, like Python. This section will introduce a few pointers, while Chap. 6 has a more in-depth examination of the different kinds of errors that can arise as you write your programs and scripts.

There are a few sources of errors that are extremely frequent that I will go over here. The first is having a missing token, which is often a missing delimiter, as those are the easiest to forget. In this example, we will take a line similar to what you entered above, but remove a delimiter.

---

[8]The ubiquity of bugs is no joke. I have been programming in one way, shape, or form for over 25 years, since a high school course in BASIC. I still routinely have bugs in almost anything I write. The difference of experience isn't that you create perfect programs the first time, but that you develop better skills at interpreting error messages and the ability to address those errors.

- Type `name = name.replace("esri", "ESRI)`. Note the missing quote mark for the last string. Because the Python Window tries to be helpful, you will actually have to delete the quote mark after the Python Window automatically enters the second quote mark when you put in the first one.
- Press Enter twice.

You should have an error message like the one below, with the cryptic "SyntaxError: EOL while scanning string literal." Returning to the ideas from the introductory concepts, in which the computer is parsing a line of code by trying to find the right pattern for the line, the SyntaxError indicates that the line didn't fit any pattern Python recognizes. The other part of this message, "EOL [End Of Line] while scanning string literal" tells us that the computer reached the end of the line before it reached the end of the string—because there was no second quote mark to indicate the end of the string.

```
>>> name = name.replace("esri", "ESRI)
  File "<string>", line 1
    name = name.replace("esri", "ESRI)
                                      ^
SyntaxError: EOL while scanning string literal
```

A similar form of syntax error can come from using the wrong delimiter. For example, you can get single quotes and double quotes mixed up. In the example below, commas and periods are mixed up.

- Type `name = name,replace("esri". "ESRI")` and press Enter twice.

Unfortunately, the error message is much less helpful here. All it tells you is that you have invalid syntax, again telling us that the line of code doesn't fit any Python pattern. In this case, it is that using a period to separate items, being the wrong use of a period, is invalid syntax.

```
>>> name = name,replace("esri". "ESRI")
  File "<string>", line 1
    name = name,replace("esri". "ESRI")
                                      ^
SyntaxError: invalid syntax
```

Replacing the period with a comma gets us a bit farther, as the different error message indicates. If the error message changes, that is often—but not always—progress. Usually, exchanging one error message for another means that the first bug has been fixed, but that there were more bugs in the program. Therefore, this is progress! However, just telling us that there is a name error is not all that useful. The closest we get to useful information is that we can tell there is a problem that it figured out when it encountered the word "replace." It turns out that it thinks replace is either a variable or a function, as opposed to a method attached to the variable called `name`, and cannot find anything by that name. We have to also change the comma in between `name` and `replace` to a period for this to work.

```
>>> name = name,replace("esri", "ESRI")
Traceback (most recent call last):
  File "<string>", line 1, in <module>
NameError: name 'replace' is not defined
```

Once the change from a period to comma is done, the line works just fine because we have successfully debugged this line.

```
>>> name = name.replace("esri", "ESRI")
>>> name
'ESRI training'
```

The last of the most common sources of bugs introduced in this section is typos with respect to variable names. In the example below, the variable `name` is misspelled as `nane`.

Type `name = nane.replace("ESRI", "ArcGIS")` and press Enter.

You will see a message about a `NameError`. This time, instead of a `SyntaxError`, in which the computer couldn't figure out what Python pattern to parse the line with, the line parsed, but there is a reference to a variable that does not exist. Specifically, this is `nane`, which it tells us in the rest of the error message, with "name 'nane' is not defined." This means that we need to ensure the variable called `nane` exists, and if it does not, changing the code to match the name you want. In this case, that would mean replacing `nane` with `name`.

```
>>> name = nane.replace("ESRI", "ArcGIS")
Traceback (most recent call last):
  File "<string>", line 1, in <module>
NameError: name 'nane' is not defined
```

The following question provides three lines with errors in them:

Question 4: Please identify and fix the errors in these lines.

```
participants append("John Andrews")
participants.reverse[]
fifth = particpiants[4]
```

Task 3: Unguided Work

In the following code, taken as lines from within this exercise, complete these tasks:
1. (10 points) Using | marks, separate the lines into their constituent tokens.
2. (15 points) Identify the variables, operators, and delimiters.

```
bob = participants[3]
name.find("GIS")
price = "$100.00"
participants[2:4]
list(paid:dictionary.keys())
```

Basics of Programming: Control Structures

3

Introductory Comments: Enabling the Computer to Make Simple Decisions

In the previous chapter, you gave the computer an instruction and it carried it out. Then you gave it the next command and it carried it out. However, the power of scripting lies in many ways in the ability of computers to carry out more complex instructions, whether that is deciding which set of instructions to do, such as send a payment reminder only if a person has not yet paid for the registration, or printing a certificate of completion for every person in the list of participants.

It would be enough of a pain to have to investigate each person, one by one, by retrieving their value from the dictionary, and deciding ourselves whether to take the next step. Additionally, it would be equally cumbersome to type in each name one by one to print off the certificate, especially when we have typed in this information once already. This means that by only doing one instruction at a time, like in Chap. 2, we are not taking advantage of benefits of scripting in the first place—the computer's ability to make comparisons and decisions extremely quickly.

It so happens that Python provides ways for us to carry out each of these two scenarios. The first of these is a conditional statement, in which the computer first investigates whether something is true or false, and takes one action or another based upon that result. In Python, conditional statements take the form of if … elif … else statements. The second of these is a loop, which repeats the same task multiple times. In Python, loops have two forms in Python: for loops and while loops.

Both conditional statements and loops are more complex statements. This is the setting, alluded to in Chap. 2, where a statement can take up multiple lines of code. Even if it is more complex, it is within this complexity that we get more value from our scripts and our programs.

Task 1: Explore the Python Notebook

This chapter's set of tasks continues where the last one left off, and even if you save your work as an ArcGIS Pro project (either the .aprx project or the .ppkx packaged project), the contents of the Python Window are not preserved. That means we should start with setting up the relevant variables before continuing here. This chapter, as well as Chaps. 4 and 8, is written to use the Python Notebook, which has been added in ArcGIS Pro version 2.8, because it provides a more direct means of recording and saving the work.[1]

Go to ArcGIS Pro and start a new Python Notebook. There are a couple ways to do this. I usually go through the Python Notebook button in the Analysis ribbon. The notebook has a small set of menus at the top, along with some buttons. Under that, you should have a blank box with In [ ]: next to it. This is where our Python code will go. Enter all of the following lines into the box:

---

[1]If your version of ArcGIS Pro is 2.7 or earlier, all code in this chapter will work within the Python Window. Just enter the first group of lines in this task to ensure all the relevant variables created in Chap. 2 are accessible and skip to Task 2. To save, though, you'd need to right-click in the upper portion of the Python Window and choose "Save Transcript." This will save both the code you entered and the resulting output, including any and all error messages.

© The Author(s), under exclusive license to Springer Nature Switzerland AG 2022
J. Conley, *A Geographer's Guide to Computing Fundamentals*, Springer Textbooks in Earth Sciences, Geography and Environment, https://doi.org/10.1007/978-3-031-08498-0_3

```
price = 100
student_price = price * 0.5
instructor = None
participants = ["John Smith", "Fred Jackson", "Alice Scholl", "Robert Waters", "Eve Douglas",
"Monty Adams", "Tyler Rose"]
participants.sort()
participants.reverse()
paid_dictionary = {}
paid_dictionary["Alice Scholl"] = True
paid_dictionary["Eve Douglas"] = False
paid_dictionary["Fred Jackson"] = False
paid_dictionary["John Smith"] = True
paid_dictionary["Monty Adams"] = True
paid_dictionary["Robert Waters"] = False
paid_dictionary["Tyler Rose"] = True
```

Run this by clicking the Run button. After it runs, there should now be a 1 in the brackets, so that this is marked as code snippet #1, by `In [1]:`. There is also another new box to receive inputted code.

There were no print statements in there, so let's confirm that the variables this creates exist, by printing off the dictionary. In the second box, enter `print(paid:dictionary)` and click the Run button. Now that has become code group #2, and the box has the printed output represented below.

```
print(paid_dictionary)
```

```
{'Alice Scholl': True, 'Eve Douglas': False, 'Fred Jackson': False, 'John Smith': True, 'Monty
Adams': True, 'Robert Waters': False, 'Tyler Rose': True}
```

One advantage of having the code blocks like this is that we can more readily edit them. Click into the first block and change the `paid_dictionary` entry for Robert Waters to `True`. Click the Run button while that block is still highlighted with a larger box around it and blue at the left.

It should change from `In [1]:` to `In [3]:` reflecting that it was re-run. Now re-run the print statement to confirm the change.

```
print(paid_dictionary)
```

```
{'Alice Scholl': True, 'Eve Douglas': False, 'Fred Jackson': False, 'John Smith': True, 'Monty
Adams': True, 'Robert Waters': True, 'Tyler Rose': True}
```

Both to illustrate part of how the Python Notebook works and reset the values for the remaining tasks of this chapter, let's change Robert's status back to False.

Edit the first group, now labeled `In [3]:`, to revert Robert back to False, but do not yet click the Run button. Instead, re-run the print statement. You should see that it does not reflect the change. This is because editing the code in the box does not, by itself, execute that code. Instead, we need to edit code in the notebook entry, and then run that code for other entries to acknowledge that change. Now run the code to commit the reversion of Robert's paid status to false, and then run the print statement to confirm that it worked.

The other immediate advantage of notebooks is that it enables easier debugging, especially of the longer pieces of code as we will be using in this chapter, and later chapters, because we can edit the block of code with the bug in it, rather than have to reenter all lines of a longer conditional or loop statement to fix a single problem on one of those lines. To illustrate this, let's say we wanted to change the first person's entry to "Jonathan Smith," but mistype with parentheses instead of brackets. In the bottom, currently empty, set of code, enter this and click Run.

```
participants(0) = "Jonathan Smith"
```

You should get this error message:

```
-----------------------------------------------------------------
SyntaxError                        Traceback (most recent call last)
File J:\PValue\arcenvironmenttest\lib\ast.py, in parse:
Line 35: return compile(source, filename, mode, PyCF_ONLY_AST)

SyntaxError: can't assign to function call (<string>, line 1)
-----------------------------------------------------------------
```

To interpret this, remember that functions and methods like the `list()` function or `.lower()` method introduced in Chap. 2 used parentheses while accessing an item in a list used brackets. Running a function is called "calling" that function, and this is something that is not permitted on the left side of an assignment operator. That's what this syntax error is complaining about, and the solution is replacing the parentheses with brackets. In the Python Window, we would have to reenter that line, although the use of the up arrow key can help. Meanwhile, here, we can edit it directly.

Replace the parentheses with brackets and click Run. The error message disappeared because this code no longer produces an error message.

Task 2: Conditional Statements

We will start with a simple `if` statement, deciding whether to send a payment notice to a participant, based upon whether they have paid. Let's start with a simple flow chart (Fig. 3.1).

As you can see, there are three parts to this. First is the question that will be used for the decision, which must be an expression that evaluates to a Boolean value. Second is the set of actions to carry out if the decision is True. Third is the set of actions to carry out if the decision is False. As we construct this statement, we will replace the "send" actions in the flow chart with `print` statements, and use Eve Douglas. Start with the decision part.

- Type `if paid_dictionary["Eve Douglas"]:`
  Make sure you include the colon. Also, recall that the dictionary is set up that the expression `paid_dictionary["Eve Douglas"]` will evaluate to either `True` or `False`.

- Press Enter once.
  The entry box has expanded to be two lines tall. This is because the computer is still waiting for the rest of the statement, and has not evaluated anything yet.
  The next line or set of lines is for what happens if the decision is true. It must be indented one level, which is typically either four or eight spaces. I cannot stress the importance of the indentation too much! Conveniently for us at this moment, the Python Notebook automatically indents the line for us.

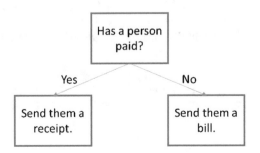

Fig. 3.1 Flow chart for conditional statement

- Type `print("send a receipt")` and press Enter once.

Lastly, we need to tell the computer what to do if the decision evaluates to false. The cue for this is through the `else` keyword. To work properly, the `else` keyword must be at the same indentation level as the `if` keyword.

- Click the mouse to move the cursor to the start of the line to undo the automatic indentation Python did for you.
- Type in `else:` (with the colon) and press Enter once.

- Lastly, type in `print("send a bill")` making sure that it lines up with the previous print line, and press Enter to finish the line of code.
  It should look like this:

```
if paid_dictionary["Eve Douglas"]:
 print("send a receipt")
else:
 print("send a bill")
```

Click the Run button, and you will see that it printed out that we need to send a bill to Eve.

```
if paid:dictionary["Eve Douglas"]:
 print("send a receipt")
else:
 print("send a bill")


send a bill
```

We might want to have a more complex flow chart, incorporating how much Eve owes, based upon whether she is a student.

In the next empty block of code, set up a `student_dictionary` variable as an empty dictionary, much like you started with `paid_dictionary`.

Then, fill it as follows:

- `student_dictionary = {}`
- `student_dictionary["Alice Scholl"] = True`
- `student_dictionary["Eve Douglas"] = False`
- `student_dictionary["Fred Jackson"] = True`
- `student_dictionary["John Smith"] = False`
- `student_dictionary["Monty Adams"] = False`
- `student_dictionary["Robert Waters"] = True`
- `student_dictionary["Tyler Rose"] = True`

Run this code.

Now our flow chart can look like this, adding in another decision of which amount to put on the bill (Fig. 3.2).

We will now revise the previous `if/else` statement to accommodate this added step. However, we will now cue the computer that we have a second condition by using a different keyword: `elif`. This is short for else if, and indicates that there is a second decision expression to run if the first decision expression evaluates to false. We still can have the `else:` clause for when all expressions are false.

- Move the cursor to the start of the `else:` line and press Enter. There should now be a blank line between `print("send a receipt")` and `else:`. It also probably decided to automatically indent the `else:` line in a failed attempt to be helpful. Move that line back to the left.

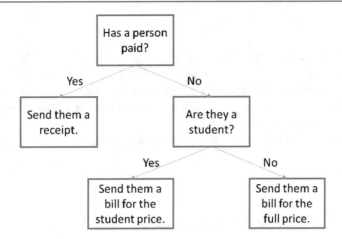

Fig. 3.2 Flow chart for if/elif/else statement

- In that blank line, type `elif student_dictionary["Eve Douglas"]:` and press Enter once.
- Now type `print("send a bill for $", student_price)` and press Enter once. This inserts the value of the variable `student_price` into the printed text.
- Specify how much to charge the nonstudent attendees to the training by revising the last line to `print("send a bill for $", price)`.

It should now look like this:

```
if paid:dictionary["Eve Douglas"]:
 print("send a receipt")
elif student_dictionary["Eve Douglas"]:
 print("send a bill for $", student_price)
else:
 print("send a bill for $", price)
```

Run the code, and the output should update to `send a bill for $ 100`.

We can do this another way, by embedding a second `if` statement inside the `else` clause of the first one. Edit the code block to be the following lines, where → represents an indentation. Note that the second `else:` must line up with its corresponding `if` at one indent in. Run the code when it is done, and you should see that the output still works.

```
if paid:dictionary["Eve Douglas"]:
→ print("send a receipt.")
else:
→ if student_dictionary["Eve Douglas"]:
→ → print("Send a bill for $", student_price)
→ else:
→ → print("Send a bill for $", price)
```

Task 3: Loops

We can carry this out for all of the participants, not just Eve, by using the other complex control statement: loops. As mentioned above, there are two types of loops in Python, `for` loops and `while` loops. Loops carry out the same lines of code several times. `For` loops repeat the code for every item in a list. We can have a simple loop that prints each participant's name. Just as the `if` statement had a line setting up the conditional statement starting with the keyword `if`, the `for` loop starts with a line setting it up with the keyword `for`. That first line has a syntax like this:

```
for [index] in [list]:
```

The index variable is a variable, which you will probably want to be a new variable, rather than one you have already declared. It is your access to each item in the list. In the first time through the loop, it refers to `list[0]`. The second time, it refers to `list[1]`. The loop continues in this manner until there are no more items in the list. As with `if` statements, the `for` line ends with a colon, and all lines of code that are to be repeated must be indented one level to the right of the `for` line. This is demonstrated in the simple loop below.

In the next empty code block, enter the following code, where again, the arrow represents an indentation:

```
for person in participants:
→ print(person)
```

Click run, and you will see each participant printed in the resulting output.

```
for person in participants:
 print(person)

Jonathan Smith
Robert Waters
Monty Adams
John Smith
Fred Jackson
Eve Douglas
Alice Scholl
```

There is another kind of loop for situations where we do not know in advance how many items we need to loop through or do not have a list ready to cycle through. It is a `while` loop, and looks similar to a `for` loop, except that where a `for` loop has a variable that iterates through a list, a `while` loop has a true/false condition. To illustrate this, we will examine another aspect of planning a GIS training, specifically what topics will be covered. To accomplish this, we will create another dictionary, this one linking the name of a topic with the amount of time (in hours) it may take to complete the training on that topic. Assume that this is a 3-day training, meaning there are 24 total hours (3 days * 8 hours per day) available. To decide what topics can be advertised, we will use a `while` loop to add topics to a list of topics until we reach 24 total hours of instruction.

Let's first create the dictionary with the following lines in a new, empty code block. Make sure you run the code to ensure this variable is accessible in other code blocks:

- `topic_dictionary = {}`
- `topic_dictionary["01_Introduction"] = 1`
- `topic_dictionary["02_Python Basics"] = 5`
- `topic_dictionary["03_Reading Python"] = 2`
- `topic_dictionary["04_ArcPy Basics"] = 2`
- `topic_dictionary["05_Debugging"] = 3`
- `topic_dictionary["06_Introduction to Datasets"] = 1`
- `topic_dictionary["07_ArcPy Cursors"] = 3`
- `topic_dictionary["08_ArcPy Geometries"] = 2`
- `topic_dictionary["09_ArcPy Rasters"] = 2`
- `topic_dictionary["10_ArcPy Map Production"] = 4`
- `topic_dictionary["11_Exceptions"] = 1`
- `topic_dictionary["12_Functions and Classes"] = 2`
- `topic_dictionary["13_Custom Tools"] = 3`
- `topic_dictionary["14_Distribution of Tools"] = 2`
- `topic_dictionary["15_Concluding Project"] = 6`

Recall from Chap. 1 the principle of initializing variables. We will need to keep track of some pieces of information as we work our way through this loop to make sure we are doing all the calculations correctly. In this case, we need to create four variables to store information for us to solve this problem. The first is the total list of topics, so that we know what topics are available for inclusion in the training. We can get this variable as the list of keys from `topic_dictionary`. Also, to make sure the inner workings of the `while` loop work properly, it should be sorted. The second variable is the list of topics that will be in the training. We will initialize this to be an empty list, which is represented by the brackets for a list with nothing in between them. Each time through the loop, we will add to this list. The third piece of information to assist our tracking is the number of hours taken by the topics in the list of training topics thus far. We will initialize it at zero, and add hours to it at the same time as we add topics to the list. The last variable is an index variable to keep track of where we are in the list of topics, which will also be initialized to zero and incremented by one each time we go through the loop.

Question 1: Please enter code to create these variables into a new, empty code block, and click Run.

Recall from above that extracting the list of keys from a dictionary does not always return them in the same order in which they were entered. This is why we need to sort the list of topics, and is why they have the numbers at the start, representing what topics are covered first, second, third, etc. Having these numbers guarantees that sorting the list will provide the appropriate ordering.

Now we are ready for our `while` loop. We can first create a flow chart of what needs to happen. Note that when the "No" part of the loop completes, it returns to the start. This means that an infinite loop can happen if the number of total hours would never exceed 24, and there was always a next topic (Fig. 3.3).

While loops start with the syntax of `while [condition]:` and the code that is indented in the following lines will keep running as long as that condition is true. This means that there is a potential problem lurking in any while loop—the infinite loop. You must ensure that the condition in the while loop will eventually become false. If it never becomes false, the loop will just keep on running and running and running, without ever stopping. If you try this in ArcGIS, you will either get a strange error or watch it run for a long time, and eventually ArcGIS will crash after running out of memory. So before you run a `while` loop, I encourage you to critically think through the code and confirm that it will end.

The while loop is as follows. Please enter it and click Run:

- `while total_hours < 24:`
- → `training_topics.append(topic_list[index])`
- → `total_hours = total_hours + topic_dictionary.get(topic_list[index])`
- → `index = index + 1`

When all is said and done, your window should not yet have any output. This is because we didn't tell the computer to also print out the names from `topic_index`. Recall that there is code to print out the names from a list in the `for` loop example above, and please revise this entry or create a new code block entry to print out the contents of the `training_`

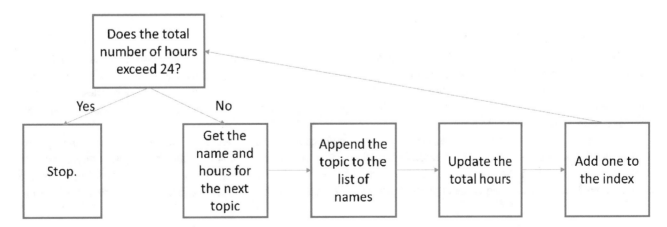

Fig. 3.3 Flow chart for `while` loop

topics list. Upon doing this, you will see that we can present the first nine topics in our training session, leaving the others to a more advanced training.

```
for item in training_topics:
 print(item)

01_Introduction
02_Python Basics
03_Reading Python
04_ArcPy Basics
05_Debugging
06_Introduction to Datasets
07_ArcPy Cursors
08_ArcPy Geometries
09_ArcPy Rasters
10_ArcPy Map Production
```

The last advantage of Python Notebooks over the Python Window is the ability to save your work more easily. There is a Save button in the Notebook Ribbon that appears when you have the Python Notebook open in ArcGIS Pro. Click it. Then save the project as Chapter3.aprx. This way, when you reopen the saved ArcGIS Pro project, your code will persist within this Notebook. However, just because it is present does not mean it has been run. You will have to re-run the blocks of code to ensure the variables they create are accessible as you continue your work.

Task 4: Parsing Complex Statements

This task looks at how the computer reads complex statements that span multiple lines. Recall from Chap. 2 that the computer parses code by looking for specific syntax patterns in the code, such as how the assignment operator requires a variable name on the left side and an expression on the right side. The computer is, in these complex statements, using similar ideas, although the patterns are simply more complicated and can cover more than one line.

Looking at a conditional statement, we have the following:

```
if paid:dictionary["Eve Douglas"]:
 → print("send a receipt.")
else:
 → if student_dictionary["Eve Douglas"]:
 → → print("Send a bill for $", student_price)
 → else:
 → → print("Send a bill for $", price)
```

It always starts with the keyword if. The computer needs this keyword to recognize that what follows is a conditional statement. The if keyword should then be followed by an expression that evaluates to a Boolean value.[2] This is why we used the dictionaries to associate the names with a series of True/False values. Because expressions can sometimes continue on and on, like x = 1 + 2 + 3 + 4 + 5, we need to give the computer a signal that the expression is done, and can be evaluated. For this reason, the line ends with a colon.

The statement(s) for the then clause, or what to do if the expression is true, immediately follow the line with the if keyword. Here, because the person has paid, a receipt should be sent. Any number of statements and therefore any number of lines can be in the section with the true clause. Therefore, like the need for a colon at the end of the Boolean expression, the computer needs some kind of signal that the set of statements for the then clause has ended. This is why there is the indenta-

---

[2]This is, strictly speaking, not an absolute requirement. It can be an integer in which positive integers are coded as True and negative integers as well as zero are coded as False. While this may be possible, that does not mean it is recommended as standard practice.

tion. That's the part of the pattern, or the invisible delimiter-like character, that denotes what is versus isn't part of the then clause statements. As mentioned above, if there is an `else:` clause, the keyword else is at the same indentation level as the keyword `if`. Any statements that are in the `else` clause are also indented a level beyond the initial keyword, to align with the statements in the then clause. If these indentation levels are incorrect, the computer will not be able to accurately figure out whether a line is or isn't inside a conditional statement. The same applies to `elif` lines in between the `if` and the `else` lines.

When the conditional statement is completed, the computer can move on to the next line(s) of code, which are aligned with the `if` and `else` keywords. The conclusion of the extra level(s) of indentation is what tells the computer that the conditional statement is over, and so the next lines must be carried out regardless of the value of the Boolean expression at the start of the if statement.

```
if [condition]:
→ statement(s)
elif [condition]:
→ statement(s)
else:
→ statement(s)
remainder of script
```

If any part of this is broken, there will be errors that can often be harder to track down than those you encountered in Chap. 2. Behind the scenes, Python numbers each of the lines in the complex statement, so that `if paid:dictionary["Eve Douglas"]:` is line number 1. Unlike lists, in this aspect of programming, we start counting the normal way, at line number 1 instead of line 0. However, if the syntax is incorrect, the computer might get partway through the remainder of the code to discover it doesn't work. The computer reports, then, the location of the error, but it reports the line on which the computer realized there was an error. For example, if you omit the colon, it will notice that the code doesn't fit the pattern when it sees a signal to move to the next line before it finds the colon. However, because of that, it discovered the error on the next line, line 2, because it only realized there was a problem when it jumped to the next line before finding the signal to officially end line 1. This means that, even if the computer reports that you have an error on line 2, the fix for it may well not be in line 2, but in line 1, or anything else that starts prior to the line which found the error.

Loops have similar patterns, especially when it comes to using tabs to demarcate blocks of code. Another similarity is that both start with keywords that indicate the purpose of the upcoming set of commands, `if`, `for`, and especially `while`. The `while` statement, like the `if` statement, must be followed by an expression that evaluates to a Boolean value, but instead of doing the true side of the flow chart once, it keeps on going until that expression is evaluated as false. Again, this means that, eventually, the expression must evaluate to false. Therefore, in one way or another, you have to ensure that the value of that expression has an opportunity to change from true to false. If you don't, and the expression always remains true, this creates an infinite loop. The only way out of an infinite loop is to force ArcGIS to close, and hope you saved everything.

```
while [condition]:
→ statement(s)
remainder of script
```

The `for` loops have a bit trickier pattern, because of the added keyword `in`. It has the pattern where the `for` keyword must be followed by a variable name, then the keyword `in`, and finally a sequence of items to iterate through. Right now, a list is the most commonplace sequence you will be using here. Like the other complex statements, the statements that fall within the loop must be indented one level.

```
for [variable] in [list]:
→ statement(s)
remainder of script
```

In this chapter, we saw conditional statements and loops as well as how to develop a flow chart to help you think through the structure of a program. Conditional statements contain if/then/elif/else portions, and even if the else clause would be empty, it can still be represented in the flow chart to reduce the possibilities of confusion. Loops come in two varieties: for and while loops, and for both, the flow chart can help structure what is inside, versus outside, the loop.

Task 5: Debugging Exercise

This is the first of several debugging exercises. While Chap. 6 goes more in depth in debugging, these debugging exercises will give you practice in tracking down common flaws in scripts which use the chapter's contents—in this case, control statements—or concepts from previous chapters. For this debugging exercise, all the errors are those in which the syntax pattern is not followed. There are three bugs in the following code, which intends to do the following.

After polling the potential participants for this GIS training, it turns out that most of them have a small amount of Python experience, which means the training can begin with topic 6 ("Introduction to Datasets") instead of the very beginning. The following code should, once corrected, print off the topics that can be included when the first five topics are skipped, starting with "06_Introduction to Datasets."

```
total_hours = 0
index = 5
training_topics = []
topic_list = list(topic_dictionary.keys())
topic_list.sort()
while total_hours < 24
→ training_topics.append(topic_list[index])
→ total_hours = total_hours + topic_dictionary.get(topic_list[index])
→ index = index + 1
for each topic in training_topics:
print(topic)
```

Task 6: Unguided Work

(1) Create a flow chart using both at least one conditional statement and at least one loop to take as input a set of potential conservation zones, and identify those which are suitable for the Golden Eagle conservation, based upon the criteria below.
   1. The area of the zone must have at least 10 acres.
   2. The zone must contain more of the forestland cover than any other land cover.
   3. The zone must be in the eastern United States.
(2) Turn that flow chart into a piece of computer code using Python, assuming you have the following variables:
   (a) zoneAreaDictionary—a dictionary in which the name of the zone is the key and the area, in acres, is the value.
   (b) zoneLandCoverDictionary—a dictionary in which the name of the zone is the key and the dominant land cover is the value.
   (c) zoneRegionDictionary—a dictionary in which the name of the zone is the key and the region in which it is found is the value.

Basics of Programming: Functions and Objects

4

Introductory Comments: Managing More Complex Code

As we build up more complex pieces of code from simple statements to complex statements, we can want blocks of code that are many statements long, but which are, nonetheless, not a full program. Using the scripting versus programming dichotomy introduced in Chap. 1, this is beginning to stretch us from the simple line-by-line scripting to a more methodologically thought-out approach to programming. If we keep to the heuristic that working in the Python Notebook or Python Window is scripting and writing a standalone Python file for a tool is programming, this is still within the scripting side of that line, but it is getting to be a fuzzy line. As we construct these more complex pieces of code, the structured reasoning processes of pseudocode and flow charts to help us figure out and manage what we want the computer to do become more critical to the success of writing the code. These blocks of code that do not form a full script are functions, which are akin to paragraphs in standard writing. Just as an essay uses paragraphs to manage the ideas and help communicate themes to the reader, a lengthy script, much less a full program, uses functions to manage the set of tasks the computer carries out.

Like variables, functions are named. Unlike variables, instead of creating a box to hold a value, functions can, perhaps, be considered like a binder of instructions. That both are named entities can create confusion, so you'll have to remember what names refer to variables and what names refer to functions. Even worse, it is possible—although very much not recommended—to have a name refer to a function, and then assign the same name to a variable, which would end up overwriting the function. All the instructions in the binder are the contents of that function, and the binder is labeled with the name, so that any time that function name gets used, the statements which are the contents of that binder are carried out. Where the comparison with paragraphs breaks down a bit is that functions allow—but do not require—input parameters and output values. Both of these are covered in more detail later in the chapter.

What is really critical for managing your code is that this means that they can be reused. While reusing paragraphs in essays is typically frowned upon as redundant writing, if not plagiarism, reusing functions is their main purpose when coding. Functions are especially powerful with the input parameters and output values mentioned above because we can reuse the function, but with different input parameters, and allow the same code to evaluate the same processes, but with different inputs to generate a different output value.

Objects add yet another layer of complexity and power to our code. They are complex data types. Whereas we have used simple types throughout, like numbers, Boolean values, and text strings, we can have far more complex types of data. These more complex types of data are objects. While objects have a wide range of flexibility in their setup, they consist largely of collections of properties and methods. Continuing the analogy of a variable as a container with the value, thinking of a simple data type as a box, we can consider an object as a crate containing variables internal to the object, which are called properties, and containing functions internal to the object, which are called methods. These are, to extend the analogy, boxes and binders within that crate, and which can refer to the other contents of the crate.

How this helps us manage our code is perhaps not nearly as straightforward as functions, as they do, indeed, add a lot of complexity. The advantage here is that objects are not simply a free-for-all where any object can have any set of properties and methods. As with anything coding-related, there is a strict structure that each object must adhere to, called its class. Each class has a set of properties that every object in that class must have, and each object has a set of methods that every object in that class must implement using the same code. That way, if we have, for example, a class for forests, every object in that class must be a forest, and have the same properties, such as the forest's size, the average age of the trees, or its dominant tree

J. Conley, *A Geographer's Guide to Computing Fundamentals*, Springer Textbooks in Earth Sciences, Geography and Environment, https://doi.org/10.1007/978-3-031-08498-0_4

species. Furthermore, every object in that class must have the same methods, such as calculating its biomass using the properties of that forest. The code for calculating the biomass will be the same for every forest, as it is part of the class blueprint, but it will use the property values specific to each individual forest to ensure the calculation is both accurate and tailored to the specific forest.

A side benefit to using functions and objects is that they also make debugging the code easier. If you had to repeat the same code, such as that biomass calculation, over and over again for each individual forest, but had a typo using a minus sign where you wanted a plus sign, you'd have to search all your code to make sure every single instance of that typo got fixed. This would be painstaking. However, if you have the task of calculating biomass neatly encapsulated within a method, then you'd have to only check that method and replace a single minus sign with a plus sign, and you know the bug is fully fixed.

Task 1: Functions

Before beginning this chapter's tasks, either open up the saved file from the end of the previous chapter if you have Python Notebooks available, or enter the following code into the Python Window if your version of ArcGIS Pro does not include the Python Notebook:

```
price = 100
student_price = price * 0.5
instructor = None
participants = ["John Smith", "Fred Jackson", "Alice Scholl", "Robert Waters", "Eve Douglas",
"Monty Adams", "Tyler Rose"]
participants.sort()
participants.reverse()
paid_dictionary = {}
paid_dictionary["Alice Scholl"] = True
paid_dictionary["Eve Douglas"] = False
paid_dictionary["Fred Jackson"] = False
paid_dictionary["John Smith"] = True
paid_dictionary["Monty Adams"] = True
paid_dictionary["Robert Waters"] = False
paid_dictionary["Tyler Rose"] = True
student_dictionary = {}
student_dictionary["Alice Scholl"] = True
student_dictionary["Eve Douglas"] = False
student_dictionary["Fred Jackson"] = True
student_dictionary["John Smith"] = False
student_dictionary["Monty Adams"] = False
student_dictionary["Robert Waters"] = True
student_dictionary["Tyler Rose"] = True
topic_dictionary = {}
topic_dictionary["01_Introduction"] = 1
topic_dictionary["02_Python Basics"] = 5
topic_dictionary["03_Reading Python"] = 2
topic_dictionary["04_ArcPy Basics"] = 2
topic_dictionary["05_Debugging"] = 3
topic_dictionary["06_Introduction to Datasets"] = 1
topic_dictionary["07_ArcPy Cursors"] = 3
topic_dictionary["08_ArcPy Geometries"] = 2
topic_dictionary["09_ArcPy Rasters"] = 2
topic_dictionary["10_ArcPy Map Production"] = 4
topic_dictionary["11_Exceptions"] = 1
topic_dictionary["12_Functions and Classes"] = 2
topic_dictionary["13_Custom Tools"] = 3
topic_dictionary["14_Distribution of Tools"] = 2
topic_dictionary["15_Concluding Project"] = 6
```

A function is a collection of statements that serve a common purpose. This common purpose or task is more complex than can be completed in a single statement, just as a paragraph is a collection of sentences that are all related to a common topic. Functions often, but not always, have one or more inputs, called parameters, and often, but again not always, have one output value.

We may want to have a function that calculates the total amount of revenue generated by the training session. The first thing to do is figure out what the inputs and output will be, if there are any. The output should be straightforward: the total amount of money that has been collected. The full set of inputs may be less obvious. The dictionary we already have of whether or not people have paid would be a good place to start. Once you know whether or not someone has paid, you need to know how much they have paid. This requires the price variable. Because of the student discount, this also requires knowing whether or not each person is a student, and what the student price is. This gives us a set of four inputs we could need.

Before we begin to write our function, we should plan it out. You might first recognize that a for loop through each of the names will be a good structure for ensuring we have checked everyone on the roster, so let's start a flow chart (Fig. 4.1).

It doesn't look like much yet, so let's next figure out the questions we need to answer for each student. We want to know if they have paid yet, and if they have paid, whether or not they get the student discount. We can then slot those questions into the flow chart, suggesting an if-else structure inside the for loop (Fig. 4.2).

Now we can fill in what to do at each of these three situations (Fig. 4.3).

Now we have the for loop filled out, but we need to know what to do before and after it. Afterward is easy: report the total revenue as output. What happens before the for loop is the initialization process to set up the loop's necessary information. In the function here, initialization takes the form of processing the input variables to make the for loop easier. One processing step is to extract a list of names from one of the dictionaries. We also need to initialize the total revenue at zero (Fig. 4.4).

With our flow chart complete, we can start our code. This involves a couple new keywords. The def keyword defines a function, and has the syntax.

```
def [function name] ([input parameter one], [input parameter 2], …):
```

For each student
in the roster

Fig. 4.1 Beginning of the function flow chart

Fig. 4.2 Fleshing out the
flow chart for our function

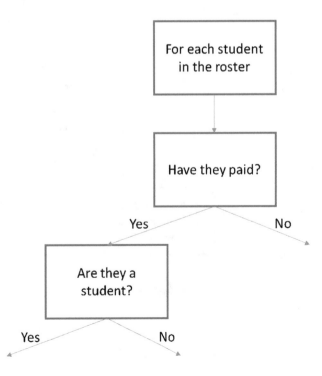

Fig. 4.3 For loop for our function

Fig. 4.4 Completed flow chart for the function

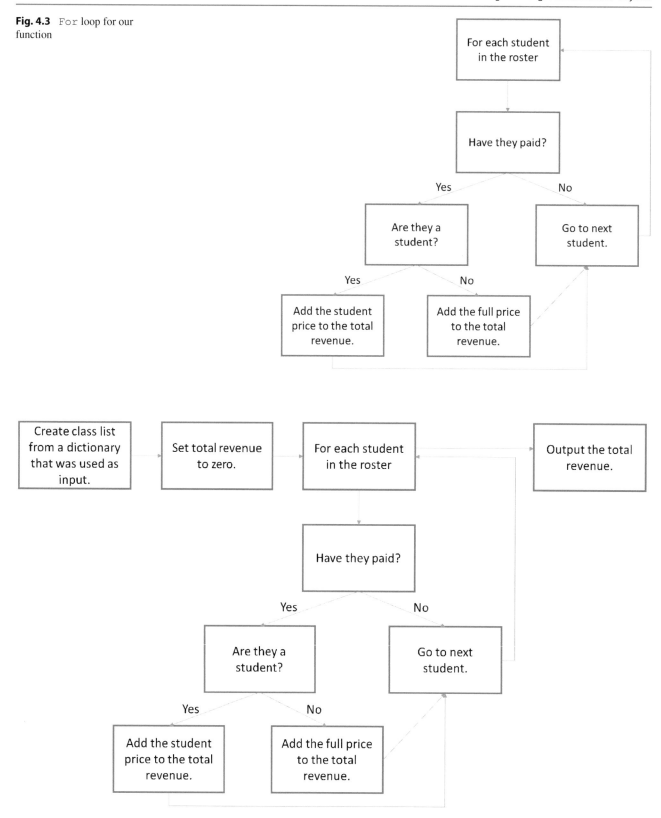

It starts with the keyword `def`, and then has the function name. Next is a tuple of parameters. If there are no parameters, you still need the empty tuple represented by a pair of parentheses with nothing between them. Lastly, there is a colon. Following the model set by control statements, any and all statements that are contained in the function must be indented one level past the `def` line.

Now we can start planning the code using comments. Comments are pieces of code that the computer ignores. While it may seem odd to include code that is deliberately ignored, it helps in the translation process between English and Computerese. If you use comments to explain what each part of the code is doing, you can make it easier to understand what your code is doing. This will also help others who you may share your code with, or even help you if you return to the code after a long absence and need the comments to help you figure out what you did. In Python, any text that follows a pound sign (#) until the end of the line is a comment. That pound sign can be at the start of a line, or in the middle of a line Let's build our code comment outline using the boxes in the flow chart.

```python
def calculate_revenue (paid_dict, student_dict, full_price, student_price):
# Create the class list.

# Set total revenue to zero.


# For each student,


# Check to see if they have paid.


# If so, check whether they were eligible for the student discount.


# If they get the discount, add the student price to the total
# revenue.  If not, add the full price to the total revenue.

# Return the total revenue as output.
```

Note a couple things about the tuple of parameters. They do not need to have the same variable names as the inputs will have. This allows you to be more flexible, and run a function with many different values. They are, though, allowed to have the same name, as `student_price` does. The parameters are given to the function as the values, rather than as references to the variable boxes themselves.[1] This means that Python makes a copy of the input and stores it in a separate variable using the name from the parameter list. While this can seem like a trivial detail, it means that if you make changes to the parameter values inside the function, those changes will not be reflected by the variable that was used as the input parameter outside the function, even if you use the same name inside and outside of the function. Put more specifically, if you try to change the student price to $60 inside this function, the change will be reflected by the calculations inside the function, but will not alter the value of the original `student_price` variable.

Now we can start to fill in the pieces of this outline. The class list is created from the set of keys from one of the input dictionaries. It does not matter which one, as they should be identical!

```python
def calculate_revenue (paid_dict, student_dict, full_price, student_price):
# Create the class list.
→ class_list = list(paid_dict.keys())
```

Next, we can add in the line setting the total revue at the start of the function.

---

[1] The more computationally advanced among you might notice this is call-by-value, not call-by-reference. There are a couple other, more obscure, parameter passing techniques: namely, call-by-name and call-by-value-reference. For the most part, just recognizing the principle of call-by-value is sufficient here.

```
# Set total revenue to zero.
→ total_revenue = 0
```

Now we can start the loop. We can use a `for` loop on the newly created `class_list` variable, and use the current item in the list to determine whether or not they have paid, through checking its value in the paid dictionary. Observe that the next line is indented twice, once to indicate it is in the function, and again to indicate that it is also in the `for` loop.

```
# For each student,
→ for student in class_list
# Check to see if they have paid.
→ → if paid_dict[student]:
```

Likewise, we can check to see if they have used the student discount with the student dictionary, and add the appropriate amount to the total revenue, indenting further as we go.

```
# If so, check whether they are eligible for the student discount.
# If they get the discount, add the student price to the total
# revenue.  If not, add the full price to the total revenue.
→ → → if student_dict[student]:
→ → → → total_revenue = total_revenue + student_price
→ → → else:
→ → → → total_revenue = total_revenue + full_price
```

While the flow chart has a step for going to the next student, we do not need to explicitly specify this step in the code, as it is already determined by how Python handles indentation within the `for` loop.

Lastly, we need to create our output, which is through the `return` keyword. All that needs done is simply returning the total revenue, as the value has been calculated along the way in the `for` loop.

```
# Return the total revenue as output.
→ return total_revenue
```

Putting all this together, we have our code. Type it all into the Python Notebook and click the Run button when you are done. (You would have to press the Enter key twice if working in the Python Window.) I've removed the comments for the purposes of making this exercise more concise. If you anticipate saving and revisiting your Python Notebook, or especially if you intend to share that Python Notebook with others, keeping the comments in could be very useful.

```
def calculate_revenue (paid_dict, student_dict, full_price, student_price):
→ class_list = list(paid_dict.keys())
→ total_revenue = 0
→ for student in class_list:
→ → if paid_dict[student]:
→ → → if student_dict[student]:
→ → → → total_revenue = total_revenue + student_price
→ → → else:
→ → → → total_revenue = total_revenue + full_price
→ return total_revenue
```

The results of doing this should be underwhelming. In fact, nothing should happen.

Question 1: Why was there no output at the completion of the block of code you just entered?

We must execute the function, which is sometimes called "calling" the function. In Python, this is done by using the function name, followed by the input parameters in parentheses. If there are no inputs, you must have an empty pair of parentheses. Furthermore, the number of input parameters needs to match the number of parameters in the function definition, and they must be presented to the function call in the same order as you defined them in the function definition.[2] Python will simply take the first value in the parentheses for the call, and put it into the first slot of the parameters' tuple, then the second value in the parentheses into the second slot of the parameters' tuple, and so on. It will not look to see if the names are similar, and will only pay attention to the order here.

Type in `calculate_revenue(paid_dictionary, student_dictionary, price, student_price)` and click Run.

Question 2 (10 points): Based upon the value you receive, how much has been collected so far?

Task 2: Objects

The other basic idea of Python we will cover in this chapter is objects. Objects are a complex kind of data type, typically more complicated than integers and Boolean values. We have, in fact, already seen some objects, because lists, tuples, dictionaries, and even strings can be considered objects. The defining characteristic of objects is that they have properties and methods attached to them. Properties are like object-specific variables, while methods are like object-specific functions. While objects in general have much flexibility in their setup, each of the objects has a specific template, called its class, to which each variable which is an instance of that class adheres. This class is a very specific blueprint of what properties and methods are associated with that object type. Creating custom classes, or custom blueprints, will be addressed in Chap. 15.

As mentioned in the introductory comments, each class has a set of properties, which are a set of values associated with each instance of this object type. For example, if we had a class of type County, it could contain the following properties (among others): population, name, state, FIPS, and geometry. These properties would be an integer for the population, a string for the name and state, another integer (or string) for FIPS, and a different type of object representing polygons for the geometry. This illustrates that the property of an object can be, and often is, another object. Every county must follow this blueprint, and contain values for these properties. Collectively, all the values of the properties is called the *state* of the object. We would then consider objects to be the same, or equal to each other, if all values within their *state* match. Within this county example, the county containing WVU might have the following state: the population is 96,189, the name is "Monongalia," the state is "WV," the FIPS is 54061, and the geometry is the polygon representing the county boundary. Even though the other 54 counties within WV have the same value for the state property, an object would only be deemed equal to Monongalia County if all properties matched Monongalia County's properties, not just one property.

Just as each class blueprint has a set of associated properties, it has a set of methods, each with specific input parameters and output values. For example, the `.append()` method is associated with lists, but not with tuples, as we have seen before, because lists can be edited while tuples are immutable. Likewise, `.lower()` is only for strings, not dictionaries or lists. The outcome of running these methods also depends entirely upon the state of the object. For example, every list has a `.sort()` method. The Python code within the `.sort()` method will be the same for every list, although the result of sorting the list will, of course, vary from list to list based upon its contents. It might even give an error message if it contains data types which cannot be sorted, such as objects.

Another delimiter is used here when working with objects, the period. It is what separates the name of the variable containing the object, and the properties and methods within that object. This period is what syntactically distinguishes a property from a variable and a method from a function.

When parsing properties, they all have the following generic syntax:

```
object_name.property_name
```

For example, if the variable `monongalia` contains the County object representing the county, you can enter `monongalia.population` and it will return the population of that county. Meanwhile, using the same property on a different county should give access to the population of that county.

---

[2]Strictly speaking, there is the possibility of optional parameters, so the function call can have fewer parameters than the function definition. However, setting up this kind of function is beyond the scope of this chapter, so we can act for the time being that they must be the same. Chapter 20, Task 3, includes how to define a function with optional parameters.

When parsing methods, they all have all the following generic syntax:

```
object_name.method_name(tuple of input parameters)
```

As with functions, even if there are no parameters, you need to have an empty pair of parentheses. This means that the string expressions like `name.lower()` and `name.find("esri")` are methods attached to the string variable called `name`. Likewise, `participants.apppend("Tyler Rose")` and `participants.sort()` are methods attached to the list variable we called `participants`.

While the structure above uses `object_name`, because of the vast majority of the time, it strictly speaking does not need a named variable containing the object. The string commands can be applied to all strings, so that the following line works:

```
"esri".upper()
```

```
'ESRI'
```

In most classes, we cannot simply create the object by putting it in quotes. So you might wonder, then, how an object gets created. This is through a specific method called a constructor. It has the same name as the class. Thus, creating an object of the type `ArcGISProject` uses the constructor method called `ArcGISProject()`. Like any other method, constructors may accept parameters, and always need the parentheses.

Task 3: Parsing Functions and Objects

Parsing of functions and objects adds another level of complexity to trying to understand what is going on in our code and troubleshoot bugs, as well. In what we had before, the code was executed line by line from start to finish, with occasional skipping over lines, thanks to conditional statements and repeating lines due to loops. Functions and objects make this more complex, and therefore, add another level of complexity to how the computer both reads and executes the code.

Specifically, by calling functions, the computer jumps from one part of a script to another. For example, if a function is defined in a script on lines 10 through 36, but is then called in line 85, the code is executed as follows: it reads lines 83, and then 84, and then when carrying out line 85, it then goes and runs lines 10 through 36. After that is done, it resumes the code starting at line 86. Therefore, while the computer may tell you an error occurred on line 29, if there are multiple calls to the same function, say on lines 50, 74, and 105, that isn't enough information to truly identify what is going on, in case the issue relates to the input parameters which could be different on lines 50, 74, 85, and 105. That additional information is necessary. To resolve this, the computer tracks an entire stack of information, each layer in the stack representing a part of how the computer got to the line it is currently executing. This is needed because just knowing how the top line of the stack, line 29 in the example, is insufficient, because there could be many ways of getting to line 29, and perhaps there is an error leading into one of the input parameters, instead of in the function code itself. In this situation, you would be told that not only was there an error in line 29, but that it occurred in the reference of line 29 that was called by the variable in line 85. This is convenient information when trying to narrow down the potential source of the problem and perhaps identify an error on line 83, which would be involved in the selection of values that get passed to the parameters of the function.

To illustrate this, let's create an error. Recall from Chap. 2 that setting the price variable to a string caused problems with code that assumes it is a number instead. Let's first create new variables to hold string representations of the prices, and then use that as inputs to the function.

```
price_str = "$100.00"
student_price_str = "$50.00"
calculate_revenue(paid_dictionary, student_dictionary, price_str, student_price_str)
```

Run this code and see what happens. You should have a type error, as before.

```
----------------------------------------------------------------------
TypeError                         Traceback (most recent call last)
In  [4]:
Line 3:     calculate_revenue(paid_dictionary, student_dictionary, price_str, student_price_str)

In  [2]:
Line 7:     total_revenue = total_revenue + student_price

TypeError: unsupported operand type(s) for +: 'int' and 'str'
----------------------------------------------------------------------
```

Within the Python Notebook, it becomes important to observe the numbers associated with each code block. Here is what I have below.

```
In [2]: def calculate_revenue (paid_dict, student_dict, full_price, student_price):
            class_list = list(paid_dict.keys())
            total_revenue = 0
            for student in class_list:
                if paid_dict[student]:
                    if student_dict[student]:
                        total_revenue = total_revenue + student_price
                    else:
                        total_revenue = total_revenue + full_price
            return total_revenue
```

```
In [3]: calculate_revenue(paid_dictionary, student_dictionary, price, student_price)
```

```
Out[3]: 300.0
```

```
In [4]: price_str = "$100.00"
        student_price_str = "$50.00"
        calculate_revenue(paid_dictionary, student_dictionary, price_str, student_price_str)

        --------------------------------------------------------------
        TypeError                         Traceback (most recent call last)
        In  [4]:
        Line 3:     calculate_revenue(paid_dictionary, student_dictionary, price_str, student_price_str)

        In  [2]:
        Line 7:     total_revenue = total_revenue + student_price

        TypeError: unsupported operand type(s) for +: 'int' and 'str'
        --------------------------------------------------------------
```

Note that block [4] is this most recent entry, and [2] is the block defining the function. It first says in the "Traceback" of the error that it is on line 3 of block [4], in which `calculate_revenue` is called. To provide more information about the nature of the error, or in case the error is in the function itself, it then says that running line 3 of block [4] required running code in block [2]. In block [2], the error was observed on line 7. That's the line that truly produced the error, as it is when the computer tried to add a number and a string. This illustrates how the computer moves from one portion of the code to another when executing functions and methods within objects, as well as reinforces the idea that just because an error was discovered in one portion of the code does not guarantee that it exists within that portion of the code.

This chapter covered both functions and objects as more complex data types that allow a greater level of complexity to the scripts that we write. Functions are covered in greater depth later in the book, Chapter 15 to be precise.

This chapter concludes the basics of Python and scripting in general. Chapter 2 introduced the building blocks of Python, as variables, operators, expressions, and statements. Chapter 3 extended this to include more complex statements, as the two types of control statements of conditionals and loops. Lastly, this chapter introduced the added complexity and depth of functions and objects.

Task 4: Debugging Exercise

As with the debugging task in Chap. 3, the following code has three errors. Please find and fix them.

The intent is to take the idea as before of identifying the topics that can fit into a training schedule, but to generalize this to any length of training, instead of just 24 instructional hours. It defines a function in which the input parameters are the dictionary of topics and hours, and the total available hours. The return value is the list of topics for the training. The following code, once debugged, will use the dictionary we had previously defined and set it up for 12 instructional hours instead of 24.

```
def identify_topics(topic_dictionary max_hours):
    index = 0
    total_hours = 0
    training_topics = []
    topic_list = list(topic_dictionary.keys())
    topic_list.sort()
    while (total_hours < max_hours) and (index < len(topic_list)):
        training_topics.append(topic_list[index])
        total_hours = total_hours + topic_dictionary.get(topic_list[index])
        index = index + 1
    return = training_topics

identify_topics(12, topic_dictionary)
```

Task 5: Non-guided Work

Please use some or all of the skills introduced and developed in Chaps. 2 through 4 to complete the following non-guided task.

Using some or all of the variables you already have as inputs, write a function that completes the following task:

Question 4: Construct a list of students who have not yet paid and return that list as the output of the function.

If you are struggling with "blank screen syndrome" and are unsure where to start, I suggest building a flow chart to begin with, so you can better plan out your code.

Reading a Python Script

5

Introductory Comments: The Value of Documentation and Comments

The previous exercises introduced basics of Python scripting. There are many scripts that you can find online, so it is common to take a script you find elsewhere and try to run it and identify what it does. Code repositories, such as GitHub, provide many scripts, but they do not exist merely as an assortment of files containing Python code. Instead, there are structures in place to help you determine what the code does. These structures are documentation and comments. Both of these help us, as the programmer, translate between English and the Computerese of Python. Documentation is information provided external to the computer code. It is typically web pages or documents that provide a detailed guide for users of the code, such as a listing of what every object does, its properties, its methods, as well as a sample piece of code calling the function or using that object. Comments, on the other hand, provide information within the code itself. It is usually more fine-grained, looking at individual blocks of code within the file and explaining them. Another way of looking at the distinction between documentation and comments is what the audience is. Documentation is for people who want to use a file, while comments are for people who want to edit a file.

The other thing being introduced in this chapter is the Integrated Development and Learning Environment (IDLE), which is the third way we will be looking at Python code in this book. Where the ArcGIS Pro Python Window is useful for working with Python code one line at a time, and the Python Notebook is useful for discrete blocks or snippets of code, the program called IDLE serves as a platform for running and editing Python programs that span an entire file (or more than one file—see Chap. 20). This introduction to IDLE will start out with a program that has already been written.

In this chapter, you will build upon the GIS training example from the previous chapters and examine a new script doing something you haven't encountered yet. You will interpret what the script does by opening it, stepping through it in IDLE, and using online help documentation to figure out commands and functions you have not yet encountered. Doing so is much like detective work, finding clues of what the script may be doing, comparing that with knowledge you already have and commands you have used, and searching for additional information to solve the problems of identifying and learning what the unfamiliar parts of the script are doing. To fit their purpose, you will examine documentation from the perspective of a user, employing existing documentation to help you decipher the script, and you will use comments to annotate the code in this script as if you were needing to share it with someone else.

Task 1: Importing a Python File into IDLE

You have been given a script called reading_python_script.py and it is up to you to figure out what it does. You also have two data files we have been using to manage the training: participants.csv and instructors.csv. We will use IDLE to run the file.

Supplementary Information: The online version contains supplementary material available at [https://doi.org/10.1007/978-3-031-08498-0_5].

- Using Windows Navigator, navigate to where you have the script saved.
- Right-click on the name and select *Edit with IDLE (ArcGIS Pro)*. Make sure you select the ArcGIS Pro option, because ArcGIS Pro and ArcGIS Desktop use different versions of Python, and this script is only guaranteed to work in Pro.

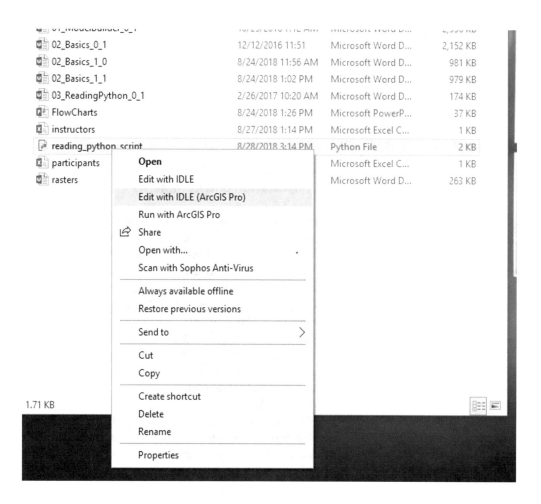

This brings up the script in a separate window. We will look at the code line by line throughout this exercise.

```
reading_python_script.py - N:\transfer2\Python\PythonBook\reading_python_script.py (3.6.2)    —    □    ✕
File  Edit  Format  Run  Options  Window  Help
import csv
participants_path = input("Enter the file path for the participants file. ")
participants_file = open(participants_path)
participants_reader = csv.reader(participants_file)
paid_dictionary = {}
student_dictionary = {}
for row in participants_reader:
    paid_dictionary[row[0]] = row[2]
    student_dictionary[row[0]] = row[1]
price = 400.0
student_price = 200.0

def calc_revenue(include_all):
    total_revenue = 0
    names = list(paid_dictionary.keys())
    for name in names:
        if (student_dictionary[name] == "TRUE"):
            if (paid_dictionary[name] == "TRUE") or include_all:
                total_revenue += student_price
        else:
            if (paid_dictionary[name] == "TRUE") or include_all:
                total_revenue += price
    return total_revenue

max_revenue = calc_revenue(True)
min_revenue = calc_revenue(False)

room_cost = 300.0
food_cost = 10*len(paid_dictionary)
min_instructor = min_revenue - room_cost - food_cost
max_instructor = max_revenue - room_cost - food_cost

instructors_path = input("Enter the file path for the instructors file. ")
instructors_file = open(instructors_path)
instructors_reader = csv.reader(instructors_file)

min_inst_list = []
max_inst_list = []

for row in instructors_reader:
                                                            Ln: 1  Col: 0
```

Let's see about getting a general overview of the script's purpose, since the script's author unhelpfully provided a non-informative file name!

- In the window with the script in it, go to the Run menu and select Run Module.
- When a prompt appears in the Python Shell window, type in the full file path to the participants file, from the drive letter (e.g., "C:\") to the extension (".csv"). For mine, it is the My Documents folder listed below. No matter where you have it, make sure that when you enter it, the forward slash (/) should be used.
- Press Enter.

- When the next prompt appears, type in the full file path for the instructor file, again from the drive letter to the extension, and press Enter.

You should then get a message of which instructors can definitely be afforded, and which are possibly affordable.

```
Enter the file path for the participants file. C:/Users/jfconley/Documents/admin/Python/partici-
pants.csv
Enter the file path for the instructors file. C:/Users/jfconley/Documents/admin/Python/instructors.
csv
We can definitely afford these instructors:
Catherine Smith
Fred Williams
We might be able to afford these instructors:
Allen Bletchley
>>>
```

This gives some indication of what the script is intended to do: help decide which instructor to use for the training.

Task 2: Use Online Python Help Documentation to Understand Part of the Script

From the start of the script, you encounter Python code we have not yet seen. The first line is `import csv`. First, notice that the word `import` is in orange. The script editor in IDLE highlights certain types of tokens in different colors to help signify the kinds of the "parts of speech," to continue the language metaphor. Keywords are in orange. We can look online to find out what the `import` keyword is doing.

Go to the Python documentation at https://docs.python.org/3/. While this brings up Python 3.9.7 documentation (at the time of writing), we want to switch over to the same version of Python that ArcGIS Pro uses, which (again, at the time of writing) is Python 3.6. Go to either the drop-down menu which currently says "3.9.7" and choose "3.6" or select the "Python 3.6 (security-fixes)" in the list of *Docs by version*.

In the *Quick search* at the top right, enter "import" and click the *Go* button. Scroll down some, and you'll see a link titled "6. Modules." Click on it to see more about modules. In practice, you might need to read through several pages of documentation before finding the right one. In this case, though, you could probably rule out those which highlighted the letters for "import" within the word "important." Modules are extensions to Python which serve specific purposes. Just like you must enable extensions like Spatial Analyst and Network Analyst within ArcGIS, you must enable extensions or modules within Python. The `import` statement is what handles that, and here we are importing a module called csv. As the name suggests, it is what allows IDLE (and Python in general) to read and write comma-separated values (.csv) files. There are many modules, including one specific for ArcGIS (called ArcPy), which we will see much more of later in the book.

Now that we have the csv module imported, we look to the next line to decipher it. It reads:

```
participants_path = input("Enter the file path for the participants
file. ")
```

First, you may notice that we have two new colors: purple and green. Based upon the syntax, we can figure out what they are. Because the purple is a name followed by parentheses, this fits the syntax for functions. Therefore, purple is for the names of functions, although it is limited to functions that are built into standard Python. (In other words, any functions you define will be black, rather than purple.) The green text starts and ends with quotation marks, telling us that it is a string.

Furthermore, you might recognize the string from when running the script. Take another look at the Python Shell and you will see that it is the prompt you had when providing the file path for the participants. That is a clue to what the `input` function does. Search again in the Python documentation for "input function" (without the quotation marks) and if you scroll down, you'll see a link for "2. Built-in Functions." Take a look at that, and you will see a table of all built-in functions at the top of the page.

Click on the `input()` entry in the table and it takes you to a description of that function. It tells you that the `input` function presents a prompt, which is the text string, and reads in whatever the user types. That input is then stored as a string in the variable on the left side of the equals sign, `participants_path` in our case.

We can also use the variable names as clues, and the name `participants_path` tells us it is the file path of that file. The next line uses the `open()` function.

```
participants_file = open(participants_path)
```

Clicking on that entry in the table informs you that it opens the file and creates it as an object of the type "file." This is distinct from the file path in the previous line. The difference between a file and a file path is important: the file path is like an address, while the file is the building at that address. My office is at 98 Beechurst Ave., Morgantown WV 26506. This is simply a piece of text telling people where it is found. It says nothing directly about the building there. It could be a house, an old building, or a new building. The building at that address can change, as the building was indeed renovated in 2007. This changed the building, just as saving a file changes the file, but the address or the file path has not changed. This is why they are separate variables. Even then, the file object is a pretty generic type, just like all you know from an address is that a building is there. Just as a file could be a Python file, an Excel file, a Word file, a PDF, etc., the building at an address could be a house, an office complex, an academic hall, an apartment building, etc. To be able to properly use the file, we need to read it. This is the job of the next line.

You will see that this next line has the same general structure as the preceding two, although the name of the function, `csv.reader`, is not purple, nor is it found in the website listing built-in functions.

```
participants_reader = csv.reader(participants_file)
```

A clue to this comes from the top line, where we imported the `csv` module. A web search for "Python csv.reader" can confirm this, where it is a function that constructs an object that can read CSV (comma-separated values) files. Note that this is still not actually reading the file. Recall that in programming, we tend to break things apart into smaller and smaller pieces, taking one incremental step at a time. As such, finding the file path, accessing the file, constructing a file reader, and reading the file line by line are all separate pieces of code. Another reason to do this is because the first parts—finding the file path and accessing the file—are generic to all file types, but the reader part is specific to CSV files. Therefore, it makes sense to have different readers for different file types, but there is no need to store or access file paths differently. Returning to the address analogy, what you do at a house versus an office complex can be different, but the addresses are interpreted consistently. Driving to 98 Beechurst Ave. is the same whether it is a house or an academic building, but you would knock on the front door of the house while entering the academic building to find the departmental office.

Lastly, once we have the file reader ready, we can carry out the task or tasks we wanted to do with the file. This is equivalent to carrying out the tasks for the building you are at, which might be selling girl scout cookies. You'd simply ask the homeowners what they want if you are at a house, but you might ask the staff in the department office to circulate a flier. Oftentimes, we will use data structures, such as lists or dictionaries, to store the contents of the file.

Task 3: Running Code Statement by Statement to Understand It

Rather than just reading line by line and looking things up in the help documentation, we can run the code one statement at a time to break it down and see what is happening at each step. We can also, at times, insert print statements or simply type in variable names to examine their values.

Open up the Python Shell by going to the *Run* menu, and selecting *Python Shell*.

This brings up a new window with a >>> prompt to enter the Python code in one statement at a time.

Since we have already looked at the first four lines, enter them in by copying them **one by one** in the code and pasting them into the Python shell. When it asks for the file path of the participants' file, please enter it.

```
>>> import csv
>>> participants_path = input("Enter the file path for the participants file. ")
```

```
Enter the file path for the participants file. C:/Users/jfconley/Documents/admin/Python/partici-
pants.csv
>>> participants_file = open(participants_path)
>>> participants_reader = csv.reader(participants_file)
```

You'll notice that there was no output, or anything printed. This is because the lines we gave either imported a module or set up a variable. If we want to see the value of a variable, we can type its name in the Python Shell and the output will be the value of that variable.

Type participants_path and press Enter. You will see the value you entered at the prompt.

```
>>> participants_path
'C:/Users/jfconley/Documents/admin/Python/participants.csv'
```

Type participants_reader and press Enter. It tells you that the participants_reader variable is an object. We introduced objects briefly at the end of the previous lab. This message tells us two pieces of information, one of which is useful. The useful information is that it is an object of the type csv.reader. We will see later how we can use this information to identify the properties this object has and the types of things it can do. The not as useful information is the hexadecimal address for the location in the computer's memory where this object is stored: 0x000001FBF524D180.

```
>>> participants_reader
<_csv.reader object at 0x000001FBF524D180>
```

Having entered these, we can move on. The next couple lines should be straightforward in setting up two dictionaries, one for whether someone is a student, and one for whether that person has paid.

```
paid_dictionary = {}
student_dictionary = {}
```

We set these up as empty dictionaries, as the empty pairs of braces can tell you.

The next loop is getting a little more complex. We would have to copy and paste the entire for statement. (This is why the start of this task said we would enter the code statement by statement instead of line by line.) Copy and paste the three lines for the for statement, but **do not** press Enter. We will have to fix some indentation. The two lines inside the for loop currently align with the keyword for. This will confuse Python.

```
>>> for row in participants_reader:
    paid_dictionary[row[0]] = row[2]
    student_dictionary[row[0]] = row[1]
```

Put the cursor before the p in paid_dictionary, press *Backspace* to move the line all the way to the left, and then press *Tab* to indent it properly. Do the same with the student_dictionary line. It should look like this when you are done.

```
>>> for row in participants_reader:
        paid_dictionary[row[0]] = row[2]
        student_dictionary[row[0]] = row[1]
```

Now you can press Enter. You will notice that the cursor is just waiting at the start of the next line. This is in case you wanted to add additional lines to the for loop.

To tell the Python Shell you are done with the for loop, press Enter again. The >>> prompt will reappear. Because nothing was printed in this loop, all you get is the prompt.

```
>>> for row in participants_reader:
        paid_dictionary[row[0]] = row[2]
        student_dictionary[row[0]] = row[1]

>>>
```

The `for` loop structure should look familiar. Notice that it is looping through items in the file reader. For CSV files, these are the rows in a spreadsheet table. The rows are imported as lists of strings. Taking a look at the CSV file, we can see that the first item is the name, the second is whether the person is a student, and the third is whether they have already paid. This means we can construct our dictionaries associating the name (`row[0]`) with the paid attribute (`row[2]`) and associating the name (`row[0]`) with their payment status (`row[1]`). Recalling how dictionaries are constructed, we can take the name and use it as the key, and the status as the value. Substituting the `row[#]` entries into the framework for adding a key-value pair to a dictionary—`dict_name[key] = value`—we get the two lines inside the for loop. Lastly, note that they are indented, telling us that they are executed for each row in the table. This is, indeed, what we want, as we would want to ensure that every participant is incorporated into our calculations. At the end of the loop, we have two dictionaries created, one linking the participant's name with whether they paid, and another linking their name with whether they are a student.

To look at the dictionaries, we can type each name into the Python Shell and press Enter.

Question 1: Looking through these dictionaries, how many participants have paid already, and how many are students?

We can now move onto the next two lines.

```
price = 400.0
student_price = 200.0
```

They are straightforward at setting up two variables, one for the general price of the training, and one for the student price of the training. A thing to note here is the use of decimal points (`200.0` instead of `200`). This tells Python to use the float data type instead of an integer. We can check the type of a variable with the built-in function `type`.

Enter `type(price)` and press Enter. You can see that it is of the type `float`.

```
>>> type(price)
<class 'float'>
>>>
```

While it doesn't make any difference for this script, there could be other situations where real numbers instead of integers are required, and this is one way to ensure that you have a real number or `float` type even if the variable is initialized with an integer value.

Task 4: Using `print` Functions to Examine a Script

The next part of the script is the function. For the Python Shell, we have to enter the function as one block or statement, rather than trying to piece it apart line by line. We can, however, make use of `print` functions to interrogate what is going on. You should recognize the basic idea of this function from the similar (but not identical) `calculate_revenue(...)` function from the previous lab. This illustrates a frequent situation when trying to decipher unfamiliar scripts. Some sections will often look familiar, so the task becomes recognizing what is the same and what is different, and the extent to which this impacts the performance of the code. To highlight this, I will show both of them below.

```
def calculate_revenue (paid_dict, student_dict, full_price, student_price):
    class_list = list(paid_dict.keys())
    total_revenue = 0
    for student in class_list:
        if paid_dict[student]:
            if student_dict[student]:
                total_revenue = total_revenue + student_price
            else:
```

```
            total_revenue = total_revenue + full_price
        return total_revenue

def calc_revenue(include_all):
    total_revenue = 0
    names = list(paid_dictionary.keys())
    for name in names:
        if (student_dictionary[name] == "TRUE"):
            if (paid_dictionary[name] == "TRUE") or include_all:
                total_revenue += student_price
        else:
            if (paid_dictionary[name] == "TRUE") or include_all:
                total_revenue += price
    return total_revenue
```

While the names are slightly different, that has no impact on the performance of the function, as long as the correct name is used. The first notable difference is the input parameters. The first one has four input parameters: the two dictionaries and the two prices. As these are all defined in the first part of the script in this lab, they are available both inside and outside the function.[1] Therefore, they do not need to be passed as input parameters. There is, however, a new input parameter, `include_all`. The name is helpful in determining what it does, but we will check shortly rather than assume its role. The order of the `total_revenue` and `names`/`class_list` lines has been flipped, but again, this has no bearing on the output of the functions. These variables can be initialized in either order.

Looking at the next line, the `for` loop cycles through the participants in either case. The `if` statements remain, although note that the order is reversed. In the first script, we wanted to tally up only those who already paid. This is the first condition because there was no point in looking at whether a participant is a student if they hadn't already paid. The second script is different, thanks to the `include_all` parameter. You can notice that whether or not a participant is a student, we check whether they paid. This is because with the `include_all` parameter, we add their price to the total whether or not they paid. This check is carried out through the `or` keyword, which is introduced in the second `if` statement. The lines inside the `if` statement will be carried out if either of the two conditions is `true`. Therefore, if `include_all` is `true`, then everyone's price will be added, while only those who have already paid will be included if `include_all` is `false`. The last difference is having a line like `total_revenue += price` instead of `total_revenue = total_revenue + full_price`. To see what the += operator does, we can go back to the Python documentation. A search for "+=" will fail, as the search process wants words, not just mathematical or punctuation characters. Search for "operators," choose the link labeled "2. Lexical analysis," and scroll to the bottom. You will see lists of operators and delimiters.

You'll see += within the delimiters section, although it functions more like an operator than a delimiter, as their text indicates. This, however, does not tell you what it does. It is, in a sense, shorthand. It needs two numeric variables, and tells the computer to take the value of what is on the right of the operator (`price` in this case), add it to the value of what is on the left (`total_revenue` here), and store that sum in the variable on the left (again, `total_revenue`). So `total_revenue += price` is equivalent to `total_revenue = total_revenue + price`. This sort of statement is so common that shorthand versions were created for all the mathematical operations (+, -, *, /, etc.).

To see the behavior of the function as it goes through the `for` loop, we might want to add `print` statements next to the two lines which update the `total_revenue` variable. So in the window with the script, add

```
print (name, " is a student, and the total is now ", total_revenue)
```

after the `total_revenue += student_price` line, and add

```
print (name, " isn't a student, and the total is now ", total_revenue)
```

after the `total_revenue += price`.

---

[1] The computer science term for this is scoping. The variables created at the start of the script are available throughout and have what is called a global scope. Those created inside the function (e.g., total_revenue) are local and cannot be accessed outside the function. In other words, they are deleted as soon as the function completes its work.

Make sure the `print` statements are indented to be inside the `if` statements.

Copy the entire `calc_revenue` function and paste it into the shell **without** pressing Enter. You'll now have to delete and reinsert **all** the tabs and indentations. Do that now, taking good care to make sure each line indented the proper number of tabs, whether 1, 2, 3, or 4. It should look similar to this. Don't panic if your `print` statements continue on a second line; because they are one line in the script, the Shell is smart enough to interpret them as one line, even if the Shell has to display them as two lines.

```
>>> def calc_revenue(include_all):
        total_revenue = 0
        names = list(paid_dictionary.keys())
        for name in names:
            if (student_dictionary[name] == "TRUE"):
                if (paid_dictionary[name] == "TRUE") or include_all:
                    total_revenue += student_price
                    print (name, " is a student, and the total is now ", total_revenue)
            else:
                if (paid_dictionary[name] == "TRUE") or include_all:
                    total_revenue += price
                    print (name, " isn't a student, and the total is now ", total_revenue)
        return total_revenue
```

Press Enter twice. You should see the >>> prompt, and nothing more.

```
>>> def calc_revenue(include_all):
        total_revenue = 0
        names = list(paid_dictionary.keys())
        for name in names:
            if (student_dictionary[name] == "TRUE"):
                if (paid_dictionary[name] == "TRUE") or include_all:
                    total_revenue += student_price
                    print (name, " is a student, and the total is now ", total_revenue)
            else:
                if (paid_dictionary[name] == "TRUE") or include_all:
                    total_revenue += price
                    print (name, " isn't a student, and the total is now ", total_revenue)
        return total_revenue

>>>
```

Question 2: Why did this not print anything, even though there are print statements?

The next two lines of code should give a big hint about the answer to question 2 above. If we calculate the revenue generated when we include all participants (i.e., setting `include_all` to `true`), that tells us how much revenue we would generate once everyone pays and thus what the maximum possible revenue is. If we do not include everyone, the other clause in the `if...or` statements determines whether or not the participant's fee gets added to the total. As that other clause is whether they've paid, it represents the amount of money that has been collected already, and is thus the minimum possible revenue.

Enter the `max_revenue = calc_revenue(True)` line into the Shell and press Enter. You now see the updated total after each participant is added to the group.

```
>>> max_revenue = calc_revenue(True)
John Smith  is a student, and the total is now  200.0
Fred Jackson  isn't a student, and the total is now  600.0
Alice Scholl  isn't a student, and the total is now  1000.0
Bob Waters  isn't a student, and the total is now  1400.0
Eve Douglas  is a student, and the total is now  1600.0
Monty Adams  is a student, and the total is now  1800.0
Tyler Rose  is a student, and the total is now  2000.0
Donald Curtis  isn't a student, and the total is now  2400.0
Katherine Clinton  isn't a student, and the total is now  2800.0
Marie Lovelace  is a student, and the total is now  3000.0
David Chang  is a student, and the total is now  3200.0
Franklin Clarke  is a student, and the total is now  3400.0
Stanley Robinson  isn't a student, and the total is now  3800.0
Jane Way  isn't a student, and the total is now  4200.0
Jeff Thomason  isn't a student, and the total is now  4600.0
Janice Waldorf  is a student, and the total is now  4800.0
Kirk Wright  is a student, and the total is now  5000.0
Craig Landingham  isn't a student, and the total is now  5400.0
Claire Frankwood  is a student, and the total is now  5600.0
```

Enter the `min_revenue = calc_revenue(False)` line into the Shell and press Enter. You now see the updated total after each participant is added to the group, but you'll notice not everyone is included this time.

```
>>> min_revenue = calc_revenue(False)
John Smith  is a student, and the total is now  200.0
Fred Jackson  isn't a student, and the total is now  .600.0
Bob Waters  isn't a student, and the total is now  1000.0
Eve Douglas  is a student, and the total is now  1200.0
Tyler Rose  is a student, and the total is now  1400.0
Donald Curtis  isn't a student, and the total is now  1800.0
David Chang  is a student, and the total is now  2000.0
Stanley Robinson  isn't a student, and the total is now  2400.0
Jane Way  isn't a student, and the total is now  2800.0
Jeff Thomason  isn't a student, and the total is now  3200.0
Kirk Wright  is a student, and the total is now  3400.0
Craig Landingham  isn't a student, and the total is now  3800.0
```

We can then confirm the two values of the `max_revenue` and `min_revenue` variables by typing them individually and pressing Enter after each one. They should match the final tally of the step-by-step running totals that were printed out in the `calc_revenue` function's `for` loops.

```
>>> max_revenue
5600.0
>>> min_revenue
3800.0
```

The next line is a straightforward setting of a variable: `room_cost`. The line after that has another unfamiliar function: `len()`. Because there's no module prefix to it (like there is with `csv.reader()` belonging to the `csv` module), we can conclude that it is a built-in function. Indeed, returning to the list of built-in functions that we looked at when starting Task

2, we can see the `len()` function is there. Clicking on that link to examine this function, we can see that it reports the length, or number of items, in a sequence or collection, like a list, a tuple, or a dictionary. In this case, it is the length of the participants' list, which would be the number of participants who have expressed interest. As the variable has the name `food_cost`, we can look at this and conclude that it is calculating the food cost as ten times the number of participants, or $10 per person. Enter these lines and press Enter.

Question 3: How many participants have expressed interest?

The next two lines are similar to each other. We are creating variables called `min_instructor` and `max_instructor`, which take the relevant revenue amounts and subtract off the costs of room and food. This would tell us how much of the revenue is left over to be able to pay the instructor's fee. Enter these lines individually, press Enter, and provide the values below.

Question 4: Using these two variables, what is the range of possible amounts we can pay the instructor?

Task 5: Using the `next()` Function to Examine the Contents of a File

The next three lines should look familiar by now. They are obviously similar to the three lines used to access the file of participants, meaning these three lines are the lines used to access the file of potential instructors. Enter them into the Python Shell, pressing Enter after each line.

```
>>> instructors_path = input("Enter the file path for the instructors file. ")


Enter the file path for the instructors file. C:/Users/jfconley/Documents/admin/Python/instructors.
csv
>>> instructors_file = open(instructors_path)
>>> instructors_reader = csv.reader(instructors_file)
>>>
```

The next two lines can also be readily interpreted. Whereas the lines after accessing the participants' file set up a couple empty dictionaries that we used a for loop to populate, these two lines set up a couple empty lists, which will be filled out in the subsequent for loop. Type these two lines into the Python Shell and press Enter after each one.

```
>>> min_inst_list = []
>>> max_inst_list = []
>>>
```

This illustrates another aspect of scripting. While there are thousands of possible commands between the core of Python and its many modules, there are certain tasks and patterns that will frequently arise, based upon what we often use computers to do. One of these tasks is reading files. A common strategy, at least for files that have the form of spreadsheets, is to set up the access to the file, create data structures (e.g., lists or dictionaries) to hold the data from the file in a more usable structure, and cycle through the file with a for loop to transfer the information from the file to the data structure. As you go through this course, you may begin to recognize patterns like this that arise from time to time. Recognizing these patterns can then help you transfer your Python skills to other languages, as the code for completing this task in other languages, like Java, which is shown below. You can see the same system of first getting the file path, then creating a file object, and finally creating a reader to access the object. The only major differences are a function called `getAllValues()` which carries out the `for` loop, and the need to specify the variable's type when creating it.

```
File nameFile = new File(filePath);
FileReader nameReader = new FileReader(nameFile);
CSVParser nameCsvp = new CSVParser(nameReader);
String[][] names = nameCsvp.getAllValues();
```

Returning to our Python script, we see another `for` loop next, as the pattern described above would suggest. There is another new function in the loop: `float`. While we could use `print` statements, similar to what we did in the previous task, this task introduces another means of examining the structure of a `for` loop, at least when going through a csv reader.

Instead of typing in the for loop just yet, type in `row = next(instructors_reader)`. Press Enter. As usual when creating a variable, there is no output. To see the contents of the variable `row`, type `row` and press Enter. You will see that this gives a list with two entries, a string containing the name of the first instructor, and a string containing that instructor's price.

```
>>> row = next(instructors_reader)
>>> row
['Caine MacEwan', '5600']
>>>
```

In effect, the `for` loop keeps calling this `next()` function until there are no more rows in the file. Once we have our row, which, as I mentioned above, is a list of strings, we can carry out the inside code of the loop one line at a time. The next line creates a variable called `inst_price`. If you look ahead, you'll see that we are going to compare this `inst_price` variable against the minimum and maximum allowable amounts from question 4. We can again use the `type()` function to confirm that the revenue variables are floats, while the entry in the `row` variable is a string. In the if statement, we cannot ask Python whether the string from `row` is greater or lesser than the float in the revenue variable. That's about as useful as asking whether 42 is less than "Fred"; the question simply makes no sense, and that doesn't change if we ask whether 42 is less than "53." Even though the string "53" contains a number, it is nonetheless still a string.

```
>>> type(max_revenue)
<class 'float'>
>>> type(min_revenue)
<class 'float'>
>>> type(row[1])
<class 'str'>
>>>
```

To ensure a valid comparison, we can turn the string into a number through this `float()` function. Enter the line to create `inst_price` and press Enter. Then use the `type()` function to confirm the data type for `inst_price`.

```
>>> inst_price = float(row[1])
>>> type(inst_price)
<class 'float'>
>>>
```

The next line simply extracts the name of the instructor to store it in a variable. Type that line and press Enter.

```
>>> inst_name = row[0]
>>>
```

Now we can enter the `if`/`elif` statement as one block into the Python Shell. As usual, make sure the indentation is correct, with the `elif` line being all the way to the left, below the >>> prompt instead of the keyword `if`. Type Enter when you have completed the conditional statement.

```
>>> if inst_price < min_instructor:
        min_inst_list.append(inst_name)
elif inst_price < max_instructor:
        max_inst_list.append(inst_name)

>>>
```

You'll again notice that there is no output. Depending upon the values of max_instructor, min_instructor, and inst_price, we might or might not have appended the instructor's name to one of the two lists. Type min_inst_list to see if the instructor's name was added to it, and press Enter. It still looks like an empty list, meaning their price was not below the minimum revenue already collected, so we might not be able to afford this instructor.

```
>>> min_inst_list
[]
>>>
```

Repeat this for max_inst_list.

Question 5: Still looking at Caine MacEwan, if everyone paid, would we be able to afford this instructor?

Before entering the entire for loop into the Python Shell, we will add a print statement to illustrate another technique for investigating a script. In the for loop that cycled through the participants, we added print statements to inform us of the running total. If we don't want to print out that much information, but only want to keep track of where we are in the loop, we can put the print statement before the if/elif statement. Insert print ("Checking on instructor ", inst_name) between the line creating the inst_name variable and the if statement. Now enter the entire for loop (with this print statement) into the Python Shell. Check (and probably correct) the indentation of each line and press Enter.

```
>>> for row in instructors_reader:
        inst_price = float(row[1])
        inst_name = row[0]
        print("Checking on instructor ", inst_name)
        if inst_price < min_instructor:
                min_inst_list.append(inst_name)
        elif inst_price < max_instructor:
                max_inst_list.append(inst_name)

Checking on instructor  Catherine Smith
Checking on instructor  Sandra Bader
Checking on instructor  Allen Bletchley
Checking on instructor  Luke Stewart
Checking on instructor  Donna Jones
Checking on instructor  Hussein Yousafzai
Checking on instructor  Fred Williams
>>>
```

You might notice that the instructor we examined for the first row, Caine MacEwan, is not printed here. This is because of our use of the next() function. As stated above, the for loop means that the file reader essentially calls next() until there are no remaining items in the list. It keeps track of which row is the next one for it to access. When we simply start into the for loop, that item is the first one, or item 0. However, our use of next() moved that index to item 1, so the for loop instead started at Catherine Smith. To illustrate this, we can try to access the next row in the table, having already completed the for loop. Type in next(instructors_reader) and press Enter. You should get an error message about a stopped iteration. This message tells us that the reader has already reached the end of the file, and there is nothing else from the file to read.

```
>>> next(instructors_reader)
Traceback (most recent call last):
  File "<pyshell#54>", line 1, in <module>
    next(instructors_reader)
StopIteration
>>>
```

Furthermore, if you re-enter the for loop, even though the print statement is in there, nothing will print out, because the reader is still at the end of the file.

```
>>> for row in instructors_reader:
        inst_price = float(row[1])
        inst_name = row[0]
        print("Checking on instructor ", inst_name)
        if inst_price < min_instructor:
                min_inst_list.append(inst_name)
        elif inst_price < max_instructor:
                max_inst_list.append(inst_name)


>>>
```

To reset the reader, we can reenter a couple lines to recreate the `instructors_file` and `instructors_reader` variables. Having done this, re-running the for loop will provide us with the output we expect, including Caine MacEwan this time.

```
>>> instructors_file = open(instructors_path)
>>> instructors_reader = csv.reader(instructors_file)
>>> for row in instructors_reader:
        inst_price = float(row[1])
        inst_name = row[0]
        print("Checking on instructor ", inst_name)
        if inst_price < min_instructor:
                min_inst_list.append(inst_name)
        elif inst_price < max_instructor:
                max_inst_list.append(inst_name)


Checking on instructor  Caine MacEwan
Checking on instructor  Catherine Smith
Checking on instructor  Sandra Bader
Checking on instructor  Allen Bletchley
Checking on instructor  Luke Stewart
Checking on instructor  Donna Jones
Checking on instructor  Hussein Yousafzai
Checking on instructor  Fred Williams
>>>
```

Next in our Python file, we have a `print` statement setting up the output in the form of the following `for` loop. That loop cycles through the instructors in the `min_inst_list` variable, telling us who we can definitely afford. Enter these lines into the Python Shell to see the content of `min_inst_list`.

Repeat this for the last three lines of the script to examine `max_inst_list`.

Question 6: Which instructor(s) can we definitely afford? Which instructor(s) can we possibly afford?

Question 7: Those instructors whose price is less than the `min_instructor` value also have a price less than `max_instructor`. However, they are not in the `max_inst_list` variable, despite meeting the criterion for appending the name to that list. Why aren't they in this list?

Question 8: If you followed the instructions above precisely, you should have each name in the two lists repeated more than once. Why would this happen?

Task 6: Adding Comments to the Code

Having deciphered the code, we would probably want to add comments into it to make it more understandable for anyone else who might happen to look at it. After all, the purpose of comments is to aid in that translation between English and Computerese. There's no strict standard for how much or how little to comment the code, but each distinct idea should be noted, so that whoever may be looking at the Python file will be able to understand what it is doing. I have an example of our file commented printed below. Recall that comments start with the # character.

```python
import csv


# Create variables to access the participants file.
participants_path = input("Enter the file path for the participants file. ")
participants_file = open(participants_path)
participants_reader = csv.reader(participants_file)

# Create empty dictionaries to store the data.
paid_dictionary = {}
student_dictionary = {}

# Cycle through the reader to put data into the dictionaries.
for row in participants_reader:
    paid_dictionary[row[0]] = row[2]
    student_dictionary[row[0]] = row[1]

# Set variables for the prices.
price = 400.0
student_price = 200.0

# Define a function to calculate the revenue.  It has a boolean input # parameter.
def calc_revenue(include_all):

    # Initialize variables for the running total and the list of names
    total_revenue = 0
    names = list(paid_dictionary.keys())

    # For eacn name...
    for name in names:

        # If they are a student, add the student price to the running
        # total if they've paid or if we are including everyone.
        if (student_dictionary[name] == "TRUE"):
            if (paid_dictionary[name] == "TRUE") or include_all:
                total_revenue += student_price

        # They aren't a student, so add the regular price to the
        # running total if they've paid or if we are including
        # everyone.
        else:
            if (paid_dictionary[name] == "TRUE") or include_all:
                total_revenue += price

    # Return the running total after the for loop is complete.
    return total_revenue
```

```python
# Run the function twice, once counting everyone (True) and once only
# counting those who have paid.
max_revenue = calc_revenue(True)
min_revenue = calc_revenue(False)

# Define some cost-related variables.  The food assumes $10 per
# person.
room_cost = 300.0
food_cost = 10*len(paid_dictionary)

# The amount of money available to pay the instructor is the revenue
# minus the cost of the room and food.
min_instructor = min_revenue - room_cost - food_cost
max_instructor = max_revenue - room_cost - food_cost

# Create variables to access the instructors file.
instructors_path = input("Enter the file path for the instructors file. ")
instructors_file = open(instructors_path)
instructors_reader = csv.reader(instructors_file)

# Create empty lists to store the data.
min_inst_list = []
max_inst_list = []

# Cycle through the reader to get data into the lists.
for row in instructors_reader:

    # Extract values from the row and store them in variables.
    inst_price = float(row[1])
    inst_name = row[0]
    print("Checking on instructor ", inst_name)

    # If the instructor's price is less than what we have already
    # collected, add them to the min_inst_list
    if inst_price < min_instructor:
        min_inst_list.append(inst_name)

    # Othewrise, if the instructor's price is less than what we could
    # eventually collect, add them to the max_inst_list.
    elif inst_price < max_instructor:
        max_inst_list.append(inst_name)

#Print out the contents of min_inst_list.
print ("We can definitely afford these instructors:")
for inst in min_inst_list:
    print (inst)

#Print out the contents of max_inst_list.
print ("We might be able to afford these instructors:")
for inst in max_inst_list:
    print (inst)
```

Task 7: Non-guided Work

There is another script available to you, cryptically titled reading_python_script_2.py. Using any or all of the strategies introduced in this lab, fill in the comments where I have started and labeled them.

You will see some familiar lines of code and a few new functions and processes. Use these strategies to identify what the new functions are.

Question 9: Use these strategies to fill in the comments on the provided script.

Grading Criteria

Comments should be concise, accurate, and free from swear words.[2]

---

[2]You might be surprisingly hard-pressed to tell the difference between a sailor and a frustrated computer programmer based upon their vocabulary.

Debugging

6

Introductory Comments: The Types of Errors

As you have undoubtedly encountered thus far, the process of scripting inevitably produces many errors. As such, before moving forward, it is useful to understand different types of errors and how they may be caught and thereby corrected. Much of the skill in debugging is in one of two areas: correctly interpreting error messages and checking the output to ensure that it did what you wanted it to do.

We can divide errors into three broad categories: syntax errors, runtime errors, and logical errors. The first two will give you the red text that you quite possibly dread now when working in the Python Window. Logical errors, however, are more insidious. They are when the code provides an answer—just not the right answer!

The distinction between syntax errors and runtime errors is a bit blurred when working within the Python Window and Notebook, although it is clearer in IDLE. With syntax errors, there is a problem with the code's grammar or syntax. This could be a misspelled word or variable name, a missing delimiter, or having the incorrect indentation. Regardless of the example, it is a situation where the computer cannot read the script to even attempt to figure out what you want it to do.

With runtime errors, the script can be read, but not executed. This can occur when you have the wrong data type for an operation. For example, if you have a variable x which is a string and a variable y which is an integer, the statement $z = x - y$ can be read. The syntax or grammar is perfect. However, when the computer tries to carry out that instruction, it realizes that it cannot subtract an integer from a string and gives an error. Another example is when you try to access an item in the list which doesn't exist. If the `index` variable has the value of 15, the statement `item = my_list[index]` can be read, but will give an error message if `my_list` only has 12 items.

In practice, within the Python Window and Notebook, both syntax and runtime errors give the red text error message as soon as the computer realizes there is a problem. The difference is that while syntax errors would almost always be found in the line that generated the error, runtime errors can occur later, only once the error becomes apparent. Consider the following set of lines:

```
x = "5"
y = 4
z = x - y
```

You won't see an error message until the third line, but the correction you'd probably want is changing the first line to $x = 5$. This illustrates the idea that while Python will give you a line number as part of the error message, that doesn't guarantee the error occurs on that line. This can make runtime errors a bit trickier to track down and fix than syntax errors.

The most challenging ones, however, are logical errors. For these, the script runs start to finish without generating any error messages. The problem, though, is that the output is not what it should be. An example would be the following attempt to calculate the distance between two points:

Supplementary Information: The online version contains supplementary material available at [https://doi.org/10.1007/978-3-031-08498-0_6].

J. Conley, *A Geographer's Guide to Computing Fundamentals*, Springer Textbooks in Earth Sciences, Geography and Environment, https://doi.org/10.1007/978-3-031-08498-0_6

```
distance = math.sqrt(((x1 - x2) * (x1 - x2)) - ((y1 - y2) * (y1 - y2)))
```

This will, as long as the parameter input expression to the `math.sqrt(...)` function is positive, return a value. Because the operation in the middle is a minus sign instead of a plus sign, while it returns a value, that value is not the correct distance between the two points. Catching this would require examining the results to ensure they are correct and make sense. For example, if this incorrect formula tells you that the distance between a town in California and a town in West Virginia is 57 miles, you should clearly have a logical error, as that is impossible. Logical errors can arise even when the code itself is perfect; the user could still provide incorrect input parameters. In the reading Python exercise, you might have, at one point, entered a file path with a typographical error. This would create an error even though the code itself is working just fine. It simply could not locate the (incorrect) file path you provided.

Task 1: Debugging in the Python Notebook

I have provided a script, called debug_demo.py, which contains ten errors. We will use the following tasks to resolve those errors. Open this file in IDLE (ArcGIS Pro), and open ArcGIS Pro itself. We will use the Python Notebook for debugging in this task, but will copy and paste the code from IDLE.

Open it in IDLE and change the file path in the second line down to wherever you have saved the relevant .csv file from previous chapters. Do not change anything else yet, even if you recognize it as an error! After updating this line, go to the File menu ➔ Save.

To run the Python file, copy the code from IDLE and paste it into the Notebook cell. Click Run. You should get some red text and an error message.

```
-----------------------------------------------------------------
SyntaxError                          Traceback (most recent call last)
File J:\PValue\arcenvironmenttest\Lib\ast.py, in parse:
Line 35:    return compile(source, filename, mode, PyCF_ONLY_AST)

SyntaxError: EOL while scanning string literal (<string>, line 2)
-----------------------------------------------------------------
```

First, it tells you what line the error is on: line 2. Contrary to how we count in lists and tuples, the line numbers here start counting at 1, so the second line down is line 2. It also tells us the line with a carat at the precise spot it recognized there was a problem. This spot appears to be at the end of the line, and it gave a syntax error "EOL while scanning string literal." The EOL means "End of Line," and it means it was expecting something before it reached the end of the line, but didn't find everything it was looking for. It tells us the EOL was found while scanning the string, which suggests something is wrong with the string. Looking closely, we might find the issue: there is no closing quotation mark. Fix this in the Notebook cell and Run the cell. You should see a different syntax error.

```
-----------------------------------------------------------------
SyntaxError                          Traceback (most recent call last)
File J:\PValue\arcenvironmenttest\Lib\ast.py, in parse:
Line 35:    return compile(source, filename, mode, PyCF_ONLY_AST)

SyntaxError: invalid syntax (<string>, line 4)
-----------------------------------------------------------------
```

This counts as progress in debugging. As long as you get to a different error message, especially if it is at a later line number, this is progress! Here we have an error on line 4. An understanding of the general syntax principles for accessing functions within modules or, for that matter, a recollection of how the script worked in the previous chapter should suggest to us that the correct syntax is "`csv.reader`," not "`csv reader`".

In the Python Notebook cell, put the dot back in between `csv` and `reader`, save the script, and rerun it in the Python Window. Yet again, there's a different syntax error, this time on line 6.

```
------------------------------------------------------------------
SyntaxError                         Traceback (most recent call last)
File J:\PValue\arcenvironmenttest\Lib\ast.py, in parse:
Line 35:     return compile(source, filename, mode, PyCF_ONLY_AST)

SyntaxError: invalid syntax (<string>, line 6)
------------------------------------------------------------------
```

Recalling the syntax format for `for` statements, what is the error here?

Question 1: What is the syntax error on line 6?

Correct the error in the Notebook, resave the script, and re-run it within the Python Notebook.

```
------------------------------------------------------------------
SyntaxError                         Traceback (most recent call last)
File J:\PValue\arcenvironmenttest\Lib\ast.py, in parse:
Line 35:     return compile(source, filename, mode, PyCF_ONLY_AST)

SyntaxError: Missing parentheses in call to 'print'. Did you mean print(name_dictionary)? (<string>,
line 11)
------------------------------------------------------------------
```

This time, we have a very helpful error message. It explicitly tells us that the parentheses in the `print` statement in line 11 are missing.[1] Put the parentheses around `name_dictionary` and re-run the script in the Python Notebook.

```
------------------------------------------------------------------
NameError                           Traceback (most recent call last)
In  [7]:
Line 3:     participants_file = open(particpiants_path)

NameError: name 'particpiants_path' is not defined
------------------------------------------------------------------
```

The next error is a `NameError`, telling us that a particular variable is not defined. First off, that the error type is not `SyntaxError` means that we have moved from syntax errors to runtime errors. Hooray! One form of bugs is eradicated from our code! The line in question, line 3, had acceptable syntax or grammar, but happened to have a typographical error in the name of a variable. In other words, Python could read the line as running the `open` function on the value stored in the variable called `particpiants_path`. It wasn't until it tried to run the function that it realized that there was no such variable. Indeed, if you look closely, you'll notice that the i and p in the middle of the variable name are reversed, making it `particpiants_path` instead of `participants_path`. Most of the time, when you get a `NameError`, especially one telling you that a variable (or function) is not defined, this means there is a typo in the name in question. Correct the spelling of the variable name and re-run in the Python Notebook.

```
------------------------------------------------------------------
TypeError                           Traceback (most recent call last)
In  [8]:
Line 8:     first_name = name_list(1)

TypeError: must be str or None, not int
------------------------------------------------------------------
```

---

[1]It is this helpful because in Python version 2.x, print statements did not require parentheses. Because of this, and especially because ArcGIS Desktop uses Python 2.x, this is likely to be a common error you encounter if you are downloading a lot of existing scripts.

Now we have a `TypeError` on line 8. Type errors are what happens when a function, operator, or method is expecting one data type, but receives a different type. In this case, based upon the error message, something is expecting a string or the empty value `None`, rather than the integer it receives.

Here is line 8: `first_name = name_list(1)`.

This might seem a bit odd. The only integer is the 1. Investigating what `name_list` is might provide us with a clue. Type `name_list` into the second cell of the Notebook and run it.

```
name_list
```

```
<built-in method split of str object at 0x0000024C89380330>
```

This probably is not what you were expecting. It seems to be telling us that the type of the variable called `name_list` is itself a built-in method called `split`. Recall that methods are attached to objects, and we see that it tells us the method is attached to a string object at `0x0000024C89380330`, a location in memory that is all but certain to be different as you follow along on your computer. If the type of a variable seems to be especially unusual, there is likely to be a problem with the line in which that variable was created. Looking at how `name_list` was created in line 7, we see that it is assigned to the value of `row[0].split`. As the error message indicates, one of the string methods is called split, and this one is attached to `row[0]`. In the bottommost cell of the Notebook, type in `row[0]` and run that cell. We see that it has the value `'John  Smith'`, which is a string object, presumably at that particular memory location. This means that the variable `name_list` was assigned to the value of the method split, as attached to that object. It is not the result you get from executing that method, but the method itself! Recognizing that if we want to execute a method, it works like executing a function, answer the following question.

Question 2: What is the error?

Correct the error and re-run that cell within the Python Notebook.

If you did the appropriate correction, you will notice that the error for this message was not on line 8, but instead earlier. It was only in line 8 that Python recognized that there was a problem. Because that variable referred to a method rather than a list, trying to access the item at index 1 was instead interpreted as calling the `.split` method assigned to the variable `name_list` with the input value of 1, and the `.split` method does not accept an integer as its input parameter.

Rerunning the code in the Notebook might not seem much like progress. We corrected a type error discovered on line 8, re-run the program, and only then found yet another type error discovered on line 8. At least the message is different!

```
----------------------------------------------------------------
TypeError                          Traceback (most recent call last)
In  [12]:
Line 8:     first_name = name_list(1)

TypeError: 'list' object is not callable
----------------------------------------------------------------
```

It is telling us that a "list" object is not callable. The first clue in this is that the problem involves a list, which is what we were expecting to get instead of a method, so this is, indeed, an improvement. The list in this line is `name_list`. Calling a method or function is another term for executing a method or function. This means it thinks we are trying to execute a method or function called `name_list`, or stored in the variable by that name, but `name_list` isn't (anymore) a method or function at all, but a list. This is how the type error arises. If we compare this line against line 9, we might get a sense of what this line is supposed to do: extract a name from the `name_list` variable. This is done through the use of brackets `[]` instead of parentheses `()`. Calling a function or method, however, uses parentheses. This suggests replacing the parentheses in line 8 with brackets. Please do this and re-run in the Python Notebook.

```
----------------------------------------------------------------
TypeError                          Traceback (most recent call last)
In  [13]:
```

```
Line 10:     name_dictionary[last_name] = first_name

TypeError: list indices must be integers or slices, not str
-------------------------------------------------------------------
```

We get yet another type error. This time, at least, we are further down, on line 10, which means we are still making progress. It says list indices must be integers or slices (which I have previously called ranges), but instead it is getting a string. The one item in line 10 that follows the syntax of a list index is `name_dictionary[last_name]`, in which `last_name` is a piece of text, rather than an integer. You might be looking at this thinking there should be no problem, as a dictionary can use a key of string type, setting up a dictionary associating the value of `first_name` with the key of `last_name`. However, the type error indicates that it isn't working with a dictionary at all, but a list, which expects an index instead of a key value.

Question 3: How would you fix this error?

Correct the error in the Notebook and re-run it within the Python Notebook.

```
{'Claire': 'Frankwood'}
```

Hooray! We have output! But not so fast. It looks like our dictionary has exactly one entry, which is notably fewer than we would want, given that there are 19 participants in the file. This means we have at least one logical error in our script. If we open the participants' file, we can see that the person in the entry, Claire Frankwood, is the last row in the table. This suggests that instead of adding all items in the file, it added only the last one. Each item is to have been added as part of the `for` loop. Notice, however, that the line which adds the item to the dictionary, line 10, is at the same indentation level as the `for` line. This means it is the first line executed **after** the `for` loop is executed. We would want it to be executed as part of the `for` loop. To move it into the loop, we need to indent line 10 one level. Please do so, and re-run the script.

```
{'John': 'Smith', 'Fred': 'Jackson', 'Alice': 'Scholl', 'Bob': 'Waters', 'Eve': 'Douglas', 'Monty':
'Adams', 'Tyler': 'Rose', 'Donald': 'Curtis', 'Katherine': 'Clinton', 'Marie': 'Lovelace', 'David':
'Chang', 'Franklin': 'Clarke', 'Stanley': 'Robinson', 'Jane': 'Way', 'Jeff': 'Thomason', 'Janice':
'Waldorf', 'Kirk': 'Wright', 'Craig': 'Landingham', 'Claire': 'Frankwood'}
```

Now that we have a much fuller output which looks like the dataset in a dictionary format, it may be tempting to call it a day. However, we can try to use the dictionary, as we may want to identify the first name of the participant Wright. In the bottommost, currently blank, cell, type in `name_dictionary['Wright']` and press Enter. Instead of the response we might expect, looking at the dictionary, "Kirk," we get an error message.

```
-------------------------------------------------------------------
KeyError                              Traceback (most recent call last)
In [16]:
Line 1:     name_dictionary['Wright']

KeyError: 'Wright'
-------------------------------------------------------------------
```

This tells us that there is no key called Wright. Since line 10, which we just indented, would seem to suggest the last name is the key and the first name is the value, this would be puzzling. The solution can be evident if we look more closely at `name_list`. Type `name_list` into the blank cell and run that cell. Notice that the first item is "Claire" and the second item is "Frankwood," yet within the `for` loop, we assign `first_name` to `name_list[1]`, or the second item, and we assign `last_name` to the first item in the list. This means our dictionary is backward, with the values where we want the keys and vice versa. To resolve this, flip the 0 and 1 in those two lines inside the `for` loop. Re-run to get the final, corrected, script working. You should notice that for each of the dictionary entries, the last name is first, and the first name is second, which follows the value and key setup we want.

```
Python                                                    ?  ▾  □  ×

 File "<string>", line 1, in <module>
KeyError: 'Wright'
name_list
['Claire', 'Frankwood']
import csv
participants_path = "N:\\transfer2\\Python\\PythonBook\\participants.csv"
participants_file = open(participants_path)
participants_reader = csv.reader(participants_file)
name_dictionary = {}
for row in participants_reader:
    name_list = row[0].split()
    first_name = name_list[0]
    last_name = name_list[1]
    name_dictionary[last_name] = first_name
print (name_dictionary)
{'Smith': 'John', 'Jackson': 'Fred', 'Scholl': 'Alice', 'Waters': 'Bob', 'Douglas':
'Eve', 'Adams': 'Monty', 'Rose': 'Tyler', 'Curtis': 'Donald', 'Clinton':
'Katherine', 'Lovelace': 'Marie', 'Chang': 'David', 'Clarke': 'Franklin',
'Robinson': 'Stanley', 'Way': 'Jane', 'Thomason': 'Jeff', 'Waldorf': 'Janice',
'Wright': 'Kirk', 'Landingham': 'Craig', 'Frankwood': 'Claire'}
|
```

Task 2: Using the IDLE Debugger

This next task illustrates the debugger built into IDLE. In Chap. 5, we used IDLE to examine an unfamiliar script that worked. This task uses IDLE in a similar manner, but to track down problems in the code. A debugger makes this task easier by providing a way of formally stepping through the code line by line while examining all the variables that are available at each step of the program. In Windows Explorer, navigate to idle_debugger_demo.py and right-click on it. Choose "Edit in IDLE (ArcGIS Pro)." You will then see that this code is similar to the `while` loop example from an earlier exercise, but has an error built into it.

```
topic_dictionary = {}
topic_dictionary["01_Python Basics"] = 5
topic_dictionary["02_Reading Python"] = 2
topic_dictionary["03_ArcPy Basics"] = 2
topic_dictionary["04_Debugging"] = 3
topic_dictionary["05_Introduction to Datasets"] = 1
topic_dictionary["06_ArcPy Cursors"] = 3
topic_dictionary["07_ArcPy Geometries"] = 2
topic_dictionary["08_ArcPy Rasters"] = 2
topic_dictionary["09_ArcPy Map Production"] = 4
topic_dictionary["10_Exceptions"] = 1
topic_dictionary["11_Functions and Classes"] = 2
topic_dictionary["12_Custom Tools"] = 3
topic_dictionary["13_Distribution of Tools"] = 2
topic_dictionary["14_Concluding Project"] = 6
```

```
index = 0
topic_list = list(topic_dictionary.keys())
training_topics = []
total_hours = 0
topic_list.sort()
while total_hours < 24:
    training_topics.append(topic_list[index])
    index = index + 1
    print("adding ", topic_list[index])
    print("total hours ", total_hours)
```

To use the debugger, first, go to the Run menu, and click on Python Shell. You now have an empty shell window. It is the same window that appeared when you ran the file in Chap. 5, but this opens it up without running anything yet. In the Debug menu, click Debugger. A new window has opened, labeled Debug Control.

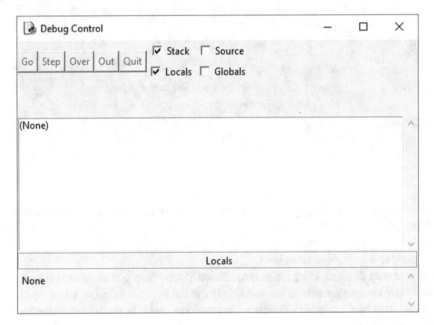

It is currently empty, but we can now go back to the code and identify where we want to make sure we step through line by line. In the code, right-click on the first line: `topic_dictionary = {}`. Click "Set Breakpoint" and that line is now highlighted in yellow. Repeat this with the line `index = 0`. When done, you should have two highlighted lines.

In the Run menu in the code window, click "Run Module," and you should see changes in the Debug Control window.

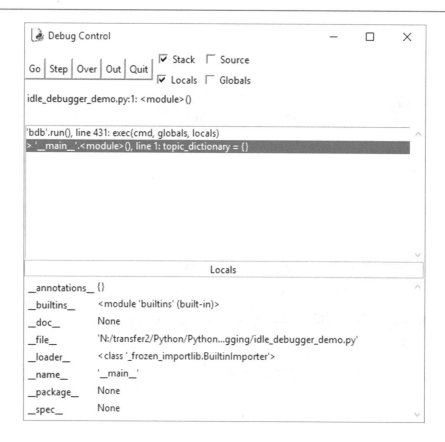

The first thing to look at is the blue highlighted line in the middle. It is telling you it is ready to execute line 1 of your code, and it tells you what that line is. It is ready to execute this line, but has not yet done so. There are also a bunch of variables whose names begin and end with underscores in the box called "Locals" at the bottom. These are variables highly specific to Python, and we do not need to worry about them here.

To execute this line, and only this line, click the button labeled "Step." You will see that it is ready to execute line 2, and carrying out line 1 has created a new variable in the variable list at the bottom: topic_dictionary, which is currently an empty dictionary. Press Step again to carry out line 2. Now you see that it is ready to execute line 3, and topic_dictionary has gained its first entry.

```
{'01_Python Basics': 5}
```

Click step again to execute line 3, and another item has been added to topic_dictionary.

```
{'01_Python Basics': 5, '02_Reading Python': 2}
```

You probably have a good sense of what is going on with these lines, and do not need to individually step through all of the lines that build the dictionary. This is why the second breakpoint was added. To fast-forward to that breakpoint, click the Go button.

Now you can see that it is ready to carry out line 17, which is where you put that breakpoint, and that the topic_dictionary variable has so many entries in it that the debugger doesn't bother to display most of them.

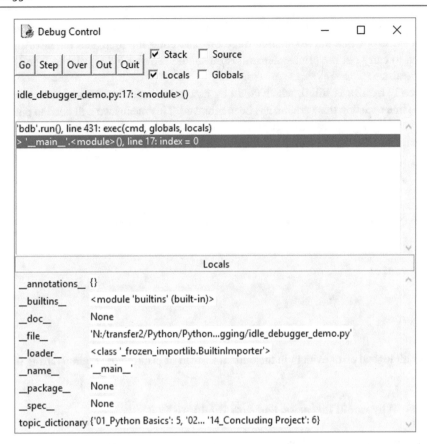

Click step to go to the next line. You'll see that the variable `index` has now been initialized to 0. Click step again to create the `topic_list` variable. Click step again to create the empty list for `training_topics`. Click step again to create `total_hours` as 0. Click step to sort the `topic_list`. You should now be ready to execute line 22, starting the `while` loop.

Click step to enter the `while` loop. Click step again to execute the first line inside the loop. You will see that `training_topics` is no longer empty, having appended the first topic. Click step again to add one to `index`. Click step again.

This now looks very different, as a lot has disappeared. Don't panic! Notice that in the code area in the middle, where you once had just the line which starts "__main__" and tells you which line you are about to execute in your script (line 25 in this case), you now have a new line below that. Line 25 in your script executes a `print` statement. It is telling you that it is currently on line 355[2] of the separate function that carries out the `print` statement. In other words, you are looking at the code for the function you called, even though it is in a different file[3]. Click step again three times. You should be back again at line 355 in the `print` function. Take a look at the Python Shell window and you should see the word "adding." This means you have carried out the first part of printing the line in line 25, which prints the string "adding" followed by the name from the `topic_list` variable.

```
>>>
[DEBUG ON]
>>>
==== RESTART: C:\Users\jfconley\Downloads\idle_debugger_demo.py ====
adding
```

Having identified what is going on, you might not want to step through all of this again. Click the Out button to finish the `print` statement. You should be back to the more familiar presentation within the Debug Control window, and ready to carry out line 26.

Since you probably don't want to go back into this `print` statement, we can go over this line with the Over button. Click the Over button to, in a sense, execute and step over the entire `print` statement and move on to the next line. Since that is

---

[2]The exact line number is subject to change in different versions of Python.

[3]The file in question is idlelib.run, to be precise.

the end of the `while` loop, it goes back to line 22, which checks the while condition. You can see that the condition is still true, as `total_hours` is 0. Click the step button three times to enter the loop, add the next topic, and update the index. Click the Over button to carry out the print statements. You should see that index is up to 2 and there are two items in the `training_topics` list.

However, the `total_hours` is still 0, which could by now raise a red flag. You might have recognized that the introduced error is that the line updating this variable has been removed. This means we will need to put it back. Having identified the problem, we do not need to continue the script until it ends. This is what the Quit button is for. Press Quit to leave the debugger and stop executing the code. Note that it did not bother to finish.

```
>>>
[DEBUG ON]
>>>
==== RESTART: C:\Users\jfconley\Downloads\idle_debugger_demo.py ====
adding  02_Reading Python
total hours  0
adding  03_ArcPy Basics
total hours  0
[DEBUG ON]
>>>
```

Also, there is a small logical error evident in the output. Instead of printing the topics starting 01 and 02, it printed out 02 and 03.

Question 4 (5 points): Why would this arise, and how would we fix it?

At this point, you can move on to task 3 without completely fixing this script by saving the change in question 4 and adding a line to update the `total_hours` variable. The purpose here is to illustrate how the debugger works.

Task 3: Non-guided Work

Using either the Python Notebook in ArcGIS Pro or the IDLE Debugger, identify and fix the eight errors in debug_task3.py.

When corrected, this will print three lists of instructors. The first two are as before, using the same code as from Chap. 5 exercises. The last is a list of those whose asking price is slightly above (within 10%) of this value, so we may be able to negotiate a rate with them that can be afforded. **All eight bugs are in this new code.** I suggest that you double-check the output with the values in the instructors.csv[4] file to identify and correct logical errors.

Your task is to correct it. As you make your corrections, add comments to the script that note the changes (e.g., #corrected misspelled variable name from vra to var).

Hints:

1. A couple of the error messages you'll get are opaque and may be hard to figure out. Don't panic!
2. Just because an error message is tied to a line doesn't guarantee the error is really on that line, as in the following example:

```
x = "5"
y = 6
z = x + y
```

The error message shows up on line 3, but the correction is to remove the quotes from line 1.

3. Just because it runs doesn't mean it runs correctly. I will check to ensure it does exactly what the description says it does.

---
[4].csv files can be opened in Excel.

Part II

Data Structures in ArcPy and Programming

The first section of this book introduced you to the basics of Python programming. By now, you might be wondering if we will ever get to the ArcGIS-specific portion of the book. This section is where it begins. Anyone who has been through even an introductory GIS course is aware of the importance of data structures in GIS and geoprocessing. The use of tables, vectors, rasters, and other data structures is inescapable for any GIS work. These structures are equally critical for programming within a GIS environment because, after all, you are almost certainly using the Python code to carry out analytical steps using these same GIS data structures within ArcGIS Pro. This section focuses upon the data structures, both as they are represented within the GIS and how Python accesses them.

Accessing GIS data structures within ArcGIS Pro requires the use of the ArcPy module. This module is introduced in Chap. 7, along with a more extended discussion of object-oriented programming, as this approach to programming is heavily used to access GIS data and tools within ArcPy. Chapter 8 more explicitly moves toward the GIS data structures within ArcPy, allowing you to determine the data type for an unknown data set. The following three chapters examine the three most common GIS data structures in more detail: rasters, tables, and vectors. The concluding chapters of this section extend the data structure discussion to look at the data structures of programming, and how Python objects provide a very structured approach to representing the information—or data—inherent in creating and manipulating maps in a GIS. These two chapters look at the symbology aspect of cartography and the layout aspect of cartography, respectively.

Introductory Comments: Objects and Classes

Before proceeding to the instructions, we will expand upon two important concepts that were introduced in Chap. 4: objects and classes. Objects were introduced in that chapter, where the ideas of properties and methods were introduced as the defining features of objects. A class is the formal structure of the properties and methods which each type of object uses.

Expanding upon the box idea from Chap. 2, an object was described in Chap. 4 as a crate containing a group of boxes, all of which are within the overarching object, as well as a set of binders representing the methods of that object. This is represented above by the variable called `aprx`, which has some of its properties shown on the left and a couple of its methods on the right, represented as simplified flow charts. Following Fig. 2.4, the boxes representing properties are color-coded by type, with String properties in green, orange representing a Map object, and gray representing a DateTime object. The class is the blueprint of what exists within each crate. For example, it specifies that each variable in the `ArcGISProject` class must have a property called `filePath` which contains a String. Likewise, it specifies that each object within this class must have a method called `listLayouts` which accepts a String as an optional input, and provides a list, and each item within that list will be a `Layout` object. Each property, like a variable, contains a value. These values are the properties of the object, and they collectively comprise the object's state. Python provides direct access to these properties, but sometimes you can only read the properties, rather than write or alter them. It is similar to the difference between lists, which can be altered after they are created, and tuples, which cannot. The distinction is that properties that are read-only can change, albeit through less direct means. Also like variables, these properties can be of any data type, including objects. In fact, this is something we will frequently encounter within ArcPy. The class blueprint enforces that every object within the class has the same set of properties, which are all of the same types, although the values of those properties can change from one object to another.

In addition to the properties, classes contain methods. These methods are similar to functions, in that they are blocks of code which have the ability to accept input parameters and provide return values. Like functions, they are not required to do so. The advantage of class-specific methods is that they can more directly use the properties, and alter the properties, of the object than having a generic function which accepts the object of its properties as an input parameter.

As this visualization approach extending the box metaphor to a crate containing multiple boxes as well as methods is cumbersome and will not scale well to more complicated classes, a visual language called Unified Modeling Language, or UML, is often employed. There are several kinds of diagrams that make up UML, and we will be focusing here on a type called a class diagram. The class diagram for Fig. 7.1 is given in Fig. 7.2.

The classes we will be working with are in a module ESRI provides, called `arcpy`. There are functions and classes built into `arcpy`, as well as a series of submodules. Each submodule has its own access. For example, the Data Access module is accessed through `arcpy.da.`, while the Mapping module is in `arcpy.mp`. As you may have observed, the dot delimiter is not just used for identifying properties and methods within classes but also for accessing classes and functions within modules. This is why we had `csv.reader` a few times in code from earlier chapters. It was accessing the `reader` class in the `csv` module. We can then access classes within those submodules by extending the dot sequence farther, and access properties and methods by extending it even farther. Thus, accessing the `ArcGISProject` class can be done through `arcpy.mp.ArcGISProject("CURRENT")`, and we can even access the file path for it with `arcpy.`

Supplementary Information: The online version contains supplementary material available at [https://doi.org/10.1007/978-3-031-08498-0_7].

Fig. 7.1 An object variable
of the class
`ArcGISProject`

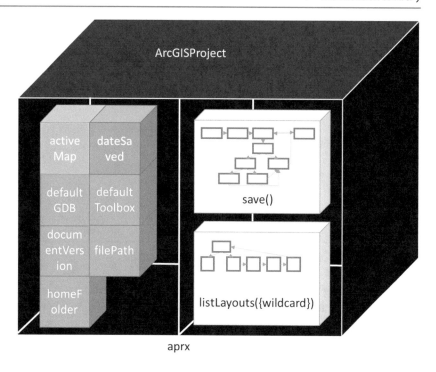

aprx

```
                    ArcGISProject
  ─────────────────────────────────────────
   activeMap: Map
   dateSaved: DateTime
   defaultGDB: String
   defaultToolbox: String
   documentVersion: String
   filePath: String
   homeFolder: String
  ─────────────────────────────────────────
   save() : void
   listLayouts({wildcard : String}) : List
```

Fig. 7.2 Class diagram representation of Fig. 7.1

`mp.ArcGISProject("CURRENT").filePath`. In theory, because properties can be objects and methods can return objects, which have their own properties and methods, we can continue to extend the dot sequence ad nauseam. However, in most situations, you would want to use a variable to hold an intermediate value, because it can be difficult to follow a long line of text punctuated with dots.

Each module, as well as the generic `arcpy` module, has functions and classes, and those classes, as described above, have their own methods and properties. The distinction between a function within a module and a method belonging to a class within that module is that the class method is accessed through a variable belonging to that class, while the function is accessed directly from the module.

As an example, if you want to print something in `arcpy`, instead of using the `print` function, which will work in the Python Window and Notebook, but not if you use a custom tool (see Chap. 14 for details), you can use the built-in `arcpy` function of `arcpy.AddMessage("message goes here")`. This function works anywhere in a program, and isn't tied to any variable. It simply prints the message. On the other hand, there is a `polygon` object, which we will see more of in Chap. 11 on geometries. The `polygon` object has properties like `isMultipart`, which tells you whether the polygon is a single unit (like West Virginia) or a series of polygons (like Hawaii). This property cannot, though, be evaluated without reference to an object of that class type. Just asking "is this polygon multipart" makes no sense if we don't know what "this polygon" refers to, we must use a variable of the type polygon to access the property `isMultipart`. Therefore, if we have a variable called `state` which is of type `polygon`, we can have an expression of `state.isMultipart` which will evaluate to `True` for Hawaii and `False` for West Virginia. Likewise, the `polygon` class has functions, which also need to go through a separate variable of the type `polygon`. One of these is `buffer`, which returns a new `polygon` object applying a buffer around the old polygon. Therefore, to find a buffer extending 50 kilometers around West Virginia, you would need something like `state.buffer(50000)` where the input parameter is the distance of the buffer expressed in map units (assuming meters for this example).

The submodules we will be working with most in this book are the Data Access, Mapping, and Spatial Analyst modules. The Data Access module, as the name suggests, is how to access, manipulate, and edit data. It is through `arcpy.da`. Mapping, which is in `arcpy.mp`, helps us create and edit the symbology of maps. The Spatial Analyst module, in `arcpy.sa`, helps us work with rasters.

Task 1: Reading ArcPy Documentation

Remembering the properties and methods of any class might seem like a lot to keep track of. This is why we have formal documentation, which serves as the English (or other language) description of a blueprint for every object within ArcPy.[1] The first class we will use is the `ArcGISProject`, and part of its documentation blueprint is reprinted below. The entire documentation is available at http://pro.arcgis.com/en/pro-app/arcpy/mapping/arcgisproject-class.htm.

The standard documentation has four main parts: a discussion, the properties of the class, the overview of the methods for the class, and lastly, a more detailed explanation of each of the methods. The discussion, which is not reprinted below, provides the general overview of what the class provides, why it exists, and how it might be used. The discussion for the ArcGISProject class points out that it is typically one of the first variables that you set when working with ArcPy scripts because it serves as the entry point for many of the other pieces of ArcPy and projects within ArcGIS Pro. The discussion also highlights that it is the means for accessing both maps and layouts, which are, after all, much of the purpose of ArcGIS in the first place.

The properties table has pretty much all the information you'd need to know about each property. The leftmost column provides the name of the property and whether it is read-only or provides write access. For example, `dateSaved` is read-only, while the `defaultGeodatabase` property can be both read and written to. This would make sense because you don't want people to be able to directly change the `dateSaved` property, because they could change it to a value which is false. It can, though, be changed indirectly, as the `save()` method will change the `dateSaved` property so that it accurately reflects when the project was last saved. The middle column, the explanation, is straightforward in telling you what that property contains. In many cases, the property name is itself self-explanatory (like `dateSaved`). The rightmost column is at least as important as the property name itself. It contains the data type of the property, whether that is a string, an object representing a date and time, or another custom object, like a `Map`.

---

[1] I won't guarantee the existence or quality of modules other than ArcPy, although many of them do have acceptable documentation. After all, if you want your module to be widely used, it is hard for that to take place if you have no documentation telling people how to use the object types in that module.

Properties

Property	Explanation	Data Type
activeMap (Read Only)	Returns the map object associated with the focused view within the application. If a layout view is active, it will return the map associated with the active map frame. 💬 Note: This property is designed to be executed from within the application using a script tool or the Python window. If a script is run outside of the application, **None** will always be returned.	Map
activeView (Read Only)	Returns a MapView or a Layout, depending on the current view. If the ArcGIS Pro project has no open views, or if the active view is something other than a map view or layout view (for example, a chart, table, Model Builder view, and so on), **None** will be returned. 💬 Note: This property is designed to be executed from within the application using a script tool or the Python window. If a script is run outside of the application, **None** will always be returned.	Object
dateSaved (Read Only)	Returns a Python **datetime** object that reports the project's last saved date.	DateTime
defaultGeodatabase (Read and Write)	The project's default geodatabase location. The string must include the geodatabase's full path and file name.	String
defaultToolbox (Read and Write)	The project's default toolbox location. The string must include the toolbox's full path and file name.	String

The next segment of the documentation is the overview table of the methods. The explanation is again a straightforward idea, but we will look at the left column in more depth. Where you saw the name of the property, you now see the name of the method, accompanied by the tuple of parameters, much like you saw with functions. The one extra thing to notice is the parameters in parentheses and curly braces ({ }). As with the function, any methods you see *must* have the tuple of parameters, even if that tuple doesn't contain any parameters. The save() function has the parentheses even though it does not require any input parameters. These parentheses are, in fact, how the Python syntax distinguishes properties from methods. Some parameters in the tuple are themselves in curly braces, like the {wildcard} parameter of the listMaps(...) method. These parameters are optional, in which case Python will use the values you provide if you choose to do so, or follow a default value if you do not provide one. The explanation column describes what the method does, and specifies what, if anything, gets returned.

Method Overview

Method	Explanation
importDocument (document_path, {include_layout}, {reuse_existing_maps})	Imports map (.mxd), globe (.3dd), and scene (.sxd) documents into an ArcGIS Pro project. It also imports the contents of map files (.mapx), layout files (.pagx), and report files (.rptx).
listBrokenDataSources ()	Returns a Python list of Layer and/or Table objects that have broken connections to their original source data for all maps in a project.
listColorRamps ({wildcard})	The listColorRamps method references color ramps available in a project.
listLayouts ({wildcard})	Returns a Python list of Layout objects in an ArcGIS project (.aprx).
listMaps ({wildcard})	Returns a Python list of Map objects in an ArcGIS project (.aprx).
listReports ({wildcard})	Returns a Python list of Report objects in an ArcGIS project (.aprx).

After looking at this overview, you might want to examine some methods in more detail. This is the last major segment of the documentation. I've pulled out three methods to highlight a few aspects. The listMaps() method illustrates the optional parameter. It has a wildcard parameter which you can use to subset the list of maps it returns by the names of those maps. The table below specifies that the data type of this parameter must be a string, although the explanation states that it does not need to be the full name. It can use an asterisk to be a bit more generic, such as "R*" referring to any and all maps whose name starts with the letter R. It also explains that if the wildcard parameter is left out, it uses the default value of None, which will indicate that it returns all maps in the project. There's a separate table for information about the return value. In this case, it informs us that the data type is a list and that we can expect the list to contain objects of the class Map.

```
listMaps ({wildcard})
```

Parameter	Explanation	Data Type
wildcard	A wildcard is based on the map name and is not case sensitive. A combination of asterisks (*) and characters can be used to help limit the resulting list. (The default value is None)	String

Return Value

Data Type	Explanation
List	A Python list of Map objects in an ArcGIS project.

Returns a Python list of Map objects in an ArcGIS project (.aprx).

The save() method has the simplest presentation here. It has both no input parameters and no return value, so both tables are absent. It has only the description of its purpose. The saveACopy(...) method illustrates how required parameters are presented without the curly braces, as it needs a file name. Note that it still only has the input parameter table; it has no return value.

Follow the link in the return value of the listMaps() method of the ArcGISProject class to open up the documentation for the Map class.

Question 1: How many properties does the Map class have?
Question 2: How many of these properties are read-only?
Question 3: How many methods does the Map class have?
Question 4: How many of these methods have optional parameters?
Question 5: Which method would you use if you found a map to be too cluttered with layers and wanted to take one or more layers out of the map?

Task 2: Accessing and Examining Basic ArcPy Classes

The tasks for this chapter will use the Python Window in ArcGIS Pro, as we are primarily looking at the code one line at a time. The first thing we will do is look at the environment workspace, as we frequently check or set that when starting a new project. There is a property of the arcpy module which is for the environment settings called arcpy.env, and by using the type() function, it can be determined to be a GPEnvironment object.[2] The workspace is a property within that GPEnvironment object.

---

[2]A search of the ArcPy documentation for "GPEnvironment" turns up no hits, oddly enough, although its structure, with a .keys() method and a .values() method, suggests to me that it is related to a dictionary, perhaps with the different settings as the keys and the values of those environment settings as the values. That this line arcpy.env["workspace"] returns the default workspace reinforces this hypothesis.

Make sure you have this chapter's data loaded into ArcGIS Pro, open the Python Window, and type `arcpy.env.workspace`. You might have noticed partway through that an autocomplete option came up. I paused after "work" and the entire property name "workspace" is visible. You can use this to identify or refresh your memory of property and method names without going to the entire documentation. Press Enter. You'll then see that the workspace is a new geodatabase created for this project. You can set it as a different geodatabase, or one you create for this project in a different location, but some of the code later on will work best if it is a geodatabase instead of a regular folder.

```
>>> arcpy.env.workspace
'C:\\Users\\jfconley\\Documents\\ArcGIS\\Projects\\Chapter7\\Chapter7.gdb'
```

As the `ArcGISProject` documentation we saw specified, that is the class that is most frequently the gateway into ArcPy. Therefore, let's create a variable containing an object of this class, referring to the current project. Conveniently, the Python Window gives us a way to get exactly that.

Type `aprx = arcpy.mp.ArcGISProject("CURRENT")` and press Enter. The "CURRENT" string is a Python Window-specific mechanism to access the currently opened project, whatever its name happens to be. If you are doing this in IDLE or another Python environment, you'd have to use the file path to specify the project you want; this is not needed in the Python Window, as it knows what is open.

To take a look briefly at the `aprx` variable, type `aprx` and press Enter.

```
>>> aprx = arcpy.mp.ArcGISProject("CURRENT")
>>> aprx
<arcpy._mp.ArcGISProject object at 0x000002AA0CEDA8D0>
```

This doesn't tell us much, although it does give us the class of the object this variable is: `ArcGISProject`. We saw a similar process previously with the `type()` function. Using it, we can see that whereas 3 is an integer, and "ArcGIS" is a string, the `aprx` variable is of the type `ArcGISProject`.

```
>>> type(aprx)
<class 'arcpy._mp.ArcGISProject'>
```

The first thing we can use `aprx` to access is the list of maps.

Type `Ohio_map = aprx.listMaps()[0]` and press Enter. Recall from the documentation that `listMaps()` provides a list of maps, even if there is only one map. This means that `aprx.listMaps()` gives us a list of one map. To access the map itself, we need to ensure the `[0]` is present at the end of the line. You can then type `Ohio_map` and press Enter to confirm that we do, indeed, have a `Map` object instead of a list.

```
>>> Ohio_map = aprx.listMaps()[0]
>>> Ohio_map
<arcpy._mp.Map object at 0x000002AA0CEDA780>
```

Now we are ready to add our data to the map. There are a few ways to do this, but one of them is to make a new layer, which automatically gets added to the map. The layer can also allow us to access much of the symbology, so this approach will be useful later on.

Type `Ohio_map.addDataFromPath("…your path…")` and press Enter. While there is a return value for this function, it is of little interest at the moment, because the primary interest in running this line is to change the map by adding the data. You will now have a map of Ohio counties on the map.

```
>>> Ohio_map.addDataFromPath("D:\\classes\\TechIssues\\data\\Ohio_ simple.shp")
<arcpy._mp.Layer object at 0x000002AA0CE89BA8>
```

Before going any further, you might want to save the script. There is no direct "Save As" menu item within the Python Window, and you have probably recognized through an earlier lab that saving the project does not, in fact, save the contents of the Python Window, even if it saves the contents of a Python Notebook. The way this saving is done is to right-click in the main part of the Python Window, and choose "Save Transcript." This brings up a "Save As" dialog box. Give your file a sensible name. To confirm that it was saved, right-click again in the main part of the Python Window and choose "Clear Transcript." This makes the Python Window empty again. To reload the transcript, right-click in the lower line, where you had been entering the code. Choose "Load Code." This will bring up a dialog box, and you can navigate to the file you just created. You should see what you just saved. You'll notice that all the input lines you had typed in are present, and any output that it gave you has been commented out. Hit enter twice to run that code. (This may well give you two "Ohio_simple" layers in the map. If so, remove one by right-clicking on the layer's name and choosing "Remove.")

We may at this point wish to see some of the properties or methods of the map. One way of accomplishing this is to type `Ohio_map.` and pause after the dot. Recall from earlier that the properties are accessed through the general syntax of `object_name.property` and the methods are similarly accessed through `object_name.method(parameters)`. Pausing after `object_name.` gets the Python Window to suggest the possible properties and methods.

Furthermore, you might notice that those with blue dots do not have parentheses. These are the properties. Those with parentheses have purple hexagons containing the letter M. These are the methods. If you start typing in a property or method name, you'll see how it gets subsetted to those which start with the character(s) you've typed in. Add an `a` after the point and pause again. You'll notice that the only methods left are those which start with a.

Using this strategy, answer question 6:

Question 6: How many *methods* does the map class have that start with the letter m?

Using either the Python Window auto-complete function or the help documentation, identify the command which can help answer question 7. Note that it is not required to start with m.

Question 7a: How many layers are in this map?

Question 7b: What was the Python statement (or statements) you used to answer this?

You should have it where the first layer is Ohio_layer.lyr, as you created it above. To examine layer objects, let's create a variable to access this layer. Regardless of the order, we can use the optional parameter to ensure we get the appropriate layer. Type:

```
>>> Ohio_layer = Ohio_map.listLayers("Ohio_simple")[0]
```

Press Enter. Again, all we know from the output of typing `Ohio_layer` and pressing enter is that it is a Layer object.

```
>>> Ohio_layer = Ohio_map.listLayers("Ohio_simple")[0]
>>> Ohio_layer
<arcpy._mp.Layer object at 0x000002AA0CF06320>
```

If we continue this process of investigating what is available by typing in `Ohio_layer.` and pause to examine the options, we might see properties called `brightness` and `contrast`. However, not all properties are relevant for all layers, because there are different kinds of layers (raster layers, vector layers, etc.) and they would have different properties. Indeed, trying `Ohio_layer.brightness` gives a bunch of red error messages. It's what is at the bottom of this message that's most important. It's the type of error that occurred and the associated explanation. You can see how it says "The attribute 'brightness' is not supported on this instance of Layer." Brightness and contrast work on rasters, but not on vectors.

```
>>> Ohio_layer.brightness
Traceback (most recent call last):
  File "c:\program files\arcgis\pro\Resources\arcpy\arcpy\arcobjects\_base.py", line 90, in _get
    return convertArcObjectToPythonObject(getattr(self._arc_object, attr_name))
AttributeError

During handling of the above exception, another exception occurred:
```

```
Traceback (most recent call last):
  File "<string>", line 1, in <module>
  File "c:\program files\arcgis\pro\Resources\arcpy\arcpy\arcobjects\_base.py", line 96, in _get
    (attr_name, self.__class__.__name__))
NameError: The attribute 'brightness' is not supported on this instance of Layer.
```

One of the more heavily used properties of the `Layer` class is its data source. Type the following lines:

```
>>> Ohio_source = Ohio_layer.dataSource
>>> Ohio_source
'D:\\classes\\TechIssues\\data\\Ohio_simple.shp'
```

This should bring up the data source. One caution: the type of this variable is a string! Confirm this using the `type` function. This means that all it does is tell the computer where to find the data, not that it contains the data itself. This is similar to the breakdown of reading a CSV file from Chap. 5, where we had separate variables and lines of code to access the file path, the file, and the object for reading the file.

If we want to look at the data that's being used, we might be tempted to try `arcpy.ListFeatureClasses()`. Type that in and press Enter. After a little while, you should see an empty list!

```
arcpy.ListFeatureClasses()
[]
```

Go back to the ArcPy help documentation. Near the top right, you should see a magnifying glass. This lets you search the documentation. Click on it and enter "ListFeatureClasses" (without any spaces). The top hit should be "ListFeatureClasses—ArcGIS Pro | Documentation." Click on that to see the function's documentation.

Between the value for `Ohio_source`, the value of `arcpy.env.workspace`, and the `ListFeatureClasses` documentation, you should be able to figure out why this returned an empty list.

Question 8: Why did `arcpy.ListFeatureClasses()` return an empty workspace?

To remedy the situation, we might want to use the following. If the Python Window moves it to a second line, that's ok, though.

```
arcpy.conversion.FeatureClassToGeodatabase(Ohio_layer, arcpy.env.workspace)
```

```
>>> arcpy.conversion.FeatureClassToGeodatabase_conversion(Ohio_layer, arcpy.env.workspace)
<Result 'C:\\Users\\jfconley\\Documents\\ArcGIS\\Projects\\Chapter7\\Chapter7.gdb'>
```

Now reenter `arcpy.ListFeatureClasses()` and press Enter. You should now see a list of one feature class.[3]

```
>>> arcpy.ListFeatureClasses()
['Ohio_simple']
```

Access it and create a new variable through the following line:

```
Ohio_fc = arcpy.ListFeatureClasses()[0]
```

---

[3]Strictly speaking, it is still a string that refers to the feature class. You can confirm this with the type() function. We will see in the upcoming exercises how to access the features themselves, rather than merely a string telling us where the features happen to reside. It so happens that for many purposes, this string is sufficient.

Task 3: Using a Tool Through ArcPy

Once we have a reference to our feature class, we would often want to apply one or more of the tools within ArcGIS to work with or analyze that feature class. The geoprocessing task we will do here is selection.

The data has records from the 2004 presidential election, so we will select those counties which voted for George W. Bush and put them into a new feature class called "Republican."

In the desktop GUI, there are several ways to accomplish this, but one of them is to go to the Analysis ribbon, and click the Tools button, to bring up the Geoprocessing pane. Do this. In the Geoprocessing pane that appeared at the right, use the search bar, currently labeled with "*Find Tools*," and search for "Select." You should then see as the top option a tool "Select (Analysis Tools)." Click it. In the tool dialog that then appears, there is a small circle with a question mark in the top right. This brings up the help file, so click it.

Scroll down in the help file and you will find a section labeled "Parameters" with two options: "Dialog" and "Python." We want the Python tab, so click that tab. This contains all the information we need to run this tool in Python. In this tab, there is a section labeled "Code sample," containing two examples, one using this tool through the Python Window and a second using this tool in a stand-alone script, such as we used in IDLE in the previous exercise.

Returning to the top of this section, we see three parameters: `in_features`, `out_feature_class`, and `where_clause`. The `in_features` requires a feature layer, so perhaps the `Ohio_layer` variable we already have would work. The output feature class will be a feature class in the environment workspace. To be explicit about this, we can create a variable for it with the following line:

```
out_fc = "Republican"
```

The last part is a SQL query. We will see SQL in greater detail when dealing with attribute data in Chap. 10. A relevant issue here is that attributes of the dataset are always included in double quotes. This means that to identify counties where the attribute Bush is greater than the attribute Kerry, we need to make sure "Bush" and "Kerry" are in double quotes. This then requires us to use single quotes to denote the start and end of the string for this query.

```
query = '"Bush" > "Kerry"'
```

We now have variables set up to run the Select tool. Enter the following line. Following the syntax at the top of the Python tab in the documentation, we have the following line:

```
arcpy.analysis.Select(Ohio_layer, out_fc, query)
```

However, note that in the code samples, a different setup is employed, which would give the following:

```
arcpy.Select_analysis("Ohio_layer", out_fc, query)
```

I recommend the former setting, as it is newer and more likely to be continually supported on a longer-term basis, while the latter approach may be deprecated, or removed from usage and support, so keeping the newer construct would be helpful. The change is part of a restructuring of submodules within ArcPy where at least some of the different toolboxes within ArcToolbox have their own module within ArcPy. The Analysis Tools toolbox contains almost all of the objects that will be associated with the tools in this toolbox. For example, each of the tools within the Extract toolbox has its own ArcPy command: `arcpy.analysis.Clip`, `arcpy.analysis.Select`, `arcpy.analysis.Split`, `arcpy.analysis.SplitByAttributes`, and `arcpy.analysis.TableSelect`.

Whether you use the old style with the toolbox indicated by _analysis at the end, or the new style with the toolbox-specific submodules, you should then get a `Result` object as a return value, and the Table of Contents at left should now have a Republican entry. Mine appeared in the exact same shade of pink as the input layer, so I suggest changing the color to make it distinguishable.

```
>>> arcpy.analysis.Select(Ohio_layer, out_fc, query)
<Result 'C:\\Users\\jfconley\\Documents\\ArcGIS\\Projects\\Chapter7\\Chapter7.gdb\\Republican'>
```

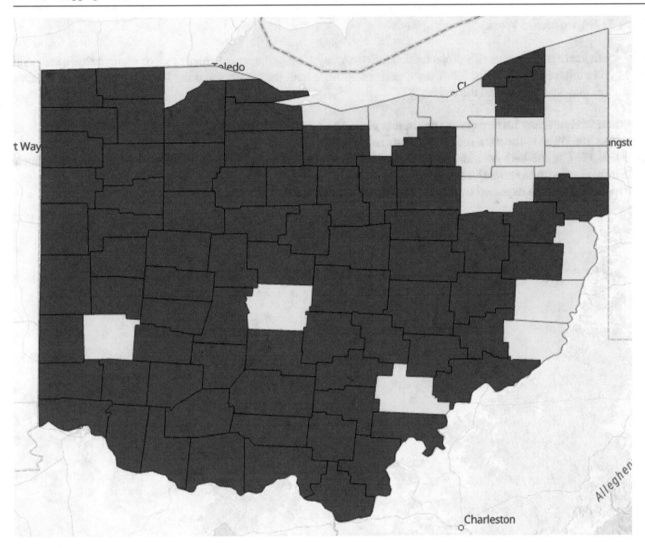

(Note: it is not required to use variables for the parameters. We could just as easily take the values of the three variables and substitute them in the `arcpy.analysis.Select` (...) line. However, I recommend using variables in such situations because if, say, you had a typo in the output feature class name, and needed to re-run the tool, you would not need to retype the query, and potentially introduce a new error if that gets a typo. You'd just need to reuse the existing `query` variable after editing the line that creates that variable. This is especially pronounced if you are reusing that query and would have to track down every single instance instead of just fixing it once.)

Task 4: Debugging Exercise

This debugging exercise provides three statements involving ArcPy code, and each statement has one error in it. Find and fix the error.

```
fields = arcpy_ListFields('Republican')
for field in fields:
    print (field.name, " is a ", fields.type)


clipped_Fc = arcpy.analysis.Clip('Republican', 'Ohio_simple',  'new_output')
```

Recall that just because it runs does not guarantee it runs accurately. There can always be logical errors.

Task 5: Non-guided Work

I have provided a different ArcGIS project, titled IntroToArcpy_Task4. Please use some or all of what you have seen thus far in this exercise to answer the following questions. Provide both the answers and the line(s) in the Python Window Transcript that was used to find the answer.

Question 9: How many feature classes are in this project?

Question 10: What is the data source for feature class #3?

Question 11: Use the help documentation to find a tool and execute it in Python to merge all of these into a single feature class with the name "Task_4."

Question 12: Does the exported transcript answer all of questions 9–11?

Introductory Comments: Superclasses, Subclasses, and Inheritance

Almost every single GIS project is going to involve data, usually a large amount of it. This chapter introduces you to working with data through ArcPy, and also introduces a programming concept called inheritance.

As we have seen, there are many variable types within Python, such as text strings, integers, real numbers, lists, dictionaries, tuples, and any kind of object. We might wish to group these into categories to help us better understand the dynamics of these different types and how they relate to one another. For example, integers and real numbers are both types of numbers. Likewise, we might see lists, tuples, and dictionaries as all collections of things. Given that, we can develop the following chart, showing these relationships among some of the basic Python data types (Fig. 8.1).

With these relationships, we know that anything that the top row can do, such as multiplying two numbers together, can also be done by the items below it in the chart. This means that you can multiply two integers together, and two real numbers together, because multiplication applies to all numbers. However, the reverse is not always true. It makes sense to get the fractional part of a real number (the part after the decimal point). However, this is not the case for an integer, as it by definition has nothing after the decimal point.

Similarly, there are some properties that apply to all collections, like asking how many values it contains. Meanwhile, you can ask for the nth item in a tuple or a list, but not a dictionary, because dictionary items are accessed through their key, rather than any indexing or ordering. We could even update our figure to reflect this added step in the hierarchy (Fig. 8.2).

Recalling the concept of classes from earlier, we can apply these concepts to objects and classes, and at the same time be more specific in our terminology. The classes at the top of an arrow are *superclasses*, while the classes at the bottom end are *subclasses*. For example, a Dictionary is a subclass of Collection, and Collection is a superclass of Dictionary. This also means that intermediate levels of a chart like this have classes that can be both. An Indexed Collection is a subclass of Collection, but a superclass of List and Tuple. The other concept here is *inheritance*. Classes of objects have methods and properties, as seen before. Any methods and properties that are implemented for a superclass are also available for all subclasses. This means that it is possible to get item #5 from an indexed collection, as this would be a method applicable to any indexed collection.[1] Because it applies for an indexed collection, it will work for both tuples and lists, as they are subclasses of the indexed collection class. However, the concept of "get item #5" doesn't work for all collections, because dictionaries aren't ordered. Therefore, "get item #5" belongs to the Indexed Collection class, and its subclasses, but not its superclass of Collection. Possibly confusing the situation further, we can, however, apply "get item" to all collections here. It's just that dictionaries and indexed collections answer the question of "which item to get" differently. Dictionaries use a key, which can be any type of object, while indexed collections must use a numerical index. This means that while using brackets [] to get an item will work with both indexed collections and dictionaries, what goes inside the brackets must be different.

These ideas apply to objects more generally, and therefore to the types of datasets within ArcPy. After all, you can load many types of data into ArcGIS, and we might be able to put them into categories based upon some of the methods or

---

[1] Strictly speaking, the [5] isn't a method, but for our purposes here, using Python data types you are familiar with, the concept of inheritance applies nonetheless.

Supplementary Information: The online version contains supplementary material available at [https://doi.org/10.1007/978-3-031-08498-0_8].

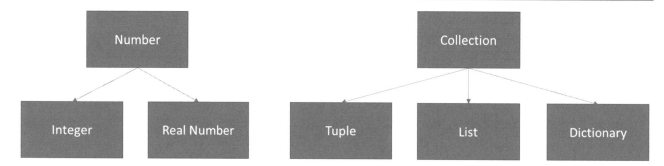

Fig. 8.1 Relationships among some Python data types

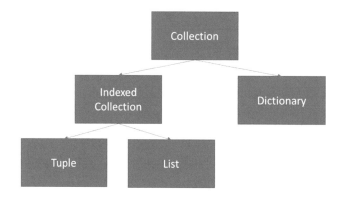

Fig. 8.2 Updated relationship among some Python data types

Fig. 8.3 Inheritance relationships among some ArcGIS dataset types

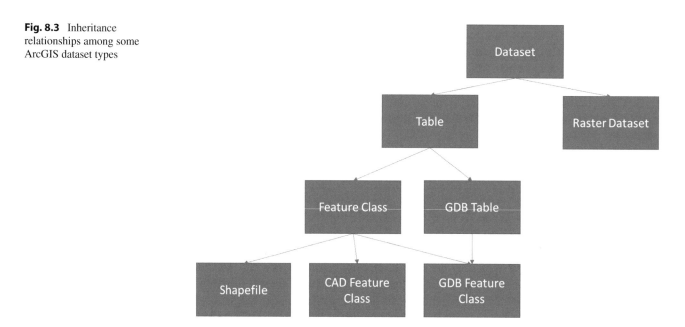

properties which can apply to them. For example, we might group all raster data types together, and all vector data types together. Figure 8.3 shows a hierarchy of some of the different ArcPy data types. This is, by no means, a complete set of all dataset types, but include the ones we may encounter most in this book. The breakdown seems to meet our commonsense conceptions of GIS data, with the initial split more or less being rasters versus vectors. I put "more or less" because of the Table superclass for all vector data, which is primarily in feature classes. Table refers to anything that has an attribute table or is a stand-alone table. For example, if you have a geodatabase (GDB) feature class of West Virginia counties, and a separate table, without any attached spatial component, of election results to join with that shapefile, both are Tables, but only the GDB feature class is a feature class. This illustrates that, under some circumstances, we can have an object that is a member of a superclass without it having to belong to any of the specific subclasses. A CSV file that you load into ArcGIS Pro will be a Table, but will not be any of the subclasses of Table, such as Feature Class or GDB Table. This doesn't always happen for superclasses, but is in this case. For example, a generic dataset cannot exist without being one of the subtypes of dataset even though a generic table can exist.

The second lesson is that a subclass can inherit from multiple superclasses. Observing the arrows, we can see that the GDB Feature Class inherits from both the GDB Table and the Feature Class. This means that any of the methods and properties associated with the GDB Table are associated with the GDB Feature Class, as is everything associated with the Feature Class. This ability to have multiple inheritance gives a great deal of flexibility to the class structure in Python in general and ArcPy in particular.

We will begin to work with GIS data using the `arcpy.Describe()` function and the resulting description object, and use those to investigate some of the inheritance structure for GIS datasets within ArcPy.

Task 1: Using arcpy.Describe

There are many scenarios where, as you are carrying out a GIS project, you are given some data which you do not know much about. Before you begin to work with it, you may wish to examine it some. While this could be done through the GUI in ArcGIS Pro, we will use ArcPy. Scripting can help here because you may consistently want to know the same pieces of information about your incoming data, ensuring a script is more useful than pointing and clicking. Moreover, if you are managing a very large project, you can be receiving more datasets than you can keep up with, thus needing the speed of the computer to help manage the flow of data.

This examination is through the Describe object, which we can access through the `arcpy.Describe()` function.[2] The Describe object is somewhat unique in that it has no methods, only properties. For that matter, it only has read-only properties—you cannot even write to the object. This may seem unusual, but recognize that for most of the properties, they are based upon the data used to create the object, and we want to ensure that the Describe object perfectly matches the data. Therefore, allowing us to write to (and therefore edit) the properties stored in the Describe object without making any changes to the corresponding data would make the Describe object incorrect. This is, to say the least, undesirable, so all properties have been marked as read-only.

Before we can use the Describe object to examine data, we need to have data to look at. I have provided a few for you on the common drive. Copy the data from the Lab6 folder to a new Lab6 folder on your N: drive (or wherever else you want to use as your workspace).

Open ArcGIS and *set the environment workspace* to the folder in which you have placed this chapter's data. You can use either the Python command or the ArcGIS GUI. There is no need to load the data to the map, although you are welcome to do so. It is, in fact, more critical that you set the environment workspace to the same place the data has been loaded, as the following lines will not work unless the workspace is pointed in the right direction.

In a new Python Notebook, run `arcpy.ListFeatureClasses()`.

Question 1: How many feature classes are in the data folder?

Now run `arcpy.ListRasters()`.

Question 2: How many raster datasets are in the data folder?

Now run `arcpy.ListTables()`.

Question 3: How many tables are in the data folder?

---

[2]If you've been observant, especially of the ArcPy documentation, you might recognize that the arcpy.Describe() function is, in fact, the constructor for an object of the type Describe. This will then be used shortly to create a variable of this class type.

To set up three lists that we will use to explore the Describe function, run the following lines in a new cell. We are working with these lists so that we can more easily keep track of what type of data we are working with, and use the Describe object more wisely in the process.

```
fcList = arcpy.ListFeatureClasses()
rsList = arcpy.ListRasters()
tbList = arcpy.ListTables()
```

Now we will set up the Describe object. It needs a single feature class (or dataset or table), rather than the list, so we need to use an index to extract a single value from the list.

```
desc = arcpy.Describe(fcList[0])
```

Now we will examine a few of the properties of the data that we can get from the Describe function. They are the dataset type, the name, the shape type, the name of field (aka attribute) #6, and the number of bands. In the Python Notebook, you may wish to put each of these in a separate cell for now, as some will generate errors and others won't. Recall that some properties are only applicable for some data types (e.g., you can't get the band count for a vector shapefile any more than you can get the fractional part of an integer). Trying to access properties that do not apply will give an error message that the method does not exist. So do not panic when at least one of these lines gives an error message. This is what you get from trying to access the band count for a vector dataset. (However, this is the same error message you would get if you spell a name wrong, so double-check your spelling and capitalization, too!)

```
desc.datasetType
'FeatureClass'

desc.name
'appalachianDevelopmentHighwaySystem_WVDOT_200903_gcs83.shp'

desc.shapeType
'Polyline'

desc.fields[6].name
'CORRIDOR_N'

desc.bandCount
---------------------------------------------------------------
AttributeError                          Traceback (most recent call last)
In [8]:
Line 5: desc.bandCount

AttributeError: DescribeData: Method bandCount does not exist
---------------------------------------------------------------
```

To show that the bandCount property exists for raster data, I'll set the desc variable to the first item in the list of rasters, and try desc.bandCount again. Now we see that the first raster in this particular workspace (which isn't the same as yours for this exercise) is of the MrSID format.

```
desc = arcpy.Describe(rsList[0])
desc.bandCount
```

1

Use the results of your code to fill in the top row of the table for question 4. Repeat this for all the other files in the three lists you created, `fcList`, `rsList`, and `tbList`. If the property is not supported (e.g., shapeType for rasters or bandCount for vectors), and you have double-checked your spelling, then you can conclude the property is not supported.

Question 4: Enter the properties here.

	datasetType	name	shapeType	fields[6].name	bandCount
fcList[0]					
fcList[1]					
fcList[2]					
fcList[3]					
rsList[0]					
rsList[1]					
tbList[0]					

Task 2: Inheritance of Properties

In the table for question 4, you should have noticed that any property supported by the table (in the row for `tbList[0]`) is also supported by the feature classes (the rows for `fcList[#]`). Likewise, a couple properties should be supported by everything. We can use the help documentation to investigate how this illustrates the ways in which these properties are inherited from superclass to subclass.

Go to https://pro.arcgis.com/en/pro-app/arcpy/functions/describe-object-properties.htm to see what is available for all Describe objects. You should see one of the properties in the table above: `name`. Because this is part of the Describe object properties, anything that can be described has a name. If you look at the list of all "Describe properties" at the left side of the window, you'll notice there are "Folder properties" among others. This means that there are things which can be described (like folders) which are not exactly datasets. Therefore, while `name` applies to truly everything, even `datasetType` doesn't apply to absolutely everything possible.

Follow the link at left to "Dataset properties." You'll notice that `datasetType` is now listed, among others, but those listed for Describe object are not. However, everything that is a dataset still has a name. This property is still accessible because it was inherited from the Describe object. The authors of the documentation did not reprint all the inherited properties, trusting that you can use the information from the Summary section to find those. For example, follow the link to "GDB FeatureClass properties." In the "Summary" section, you will see a list of all the super/sub classes between the overarching Describe object and the specific GDB Feature Class here. In this case, in addition to Describe and GDB Feature Class, all properties belonging to FeatureClass, GDB Table, Editor Tracking Dataset, Table, and Dataset classes are also inherited by the GDB FeatureClass. Therefore, while it looks a bit underwhelming through the documentation itself, where there are only four properties listed, it is important to acknowledge that with GDB FeatureClasses, we can also access the 12 properties in the Describe object, the six properties for the FeatureClass object, the ten properties for GDB Table, and so on. It is just these four which are unique to the GDB FeatureClass itself and any subclasses it may have.

Question 5 (5 points): Using those inherited from _all_ superclasses, including Describe, how many properties are available for Shapefile FeatureClasses?

Question 6 (5 points): Using those inherited from _all_ superclasses, including Describe, how many properties are available for Raster Datasets?

Task 3: Running an ArcPy Script in IDLE

You may have noticed in many of the code samples in the ArcPy documentation that there are examples for using the function, method, or object in the Python Window within ArcGIS Pro, as well as in a stand-alone script that would run in an outside environment, like IDLE. We will start here with accessing `arcpy` commands through IDLE instead of the Python Window.

For something as basic as this, where we access datasets to describe them, but make no attempt to edit them, use them for analysis, or create new data, the transition from the Python Window or Notebook to IDLE is not as great as you might fear. I introduce the necessary steps here, and you will, in the non-guided task, be writing a script that uses these ideas to operate in IDLE instead of the Python Window.

First, recall that `arcpy` is a Python module, and when working outside of the Python Window in ArcGIS Pro, we must `import` it like any other module. Therefore, you'll have to start off with an `import arcpy` statement.

Second, you might have used the GUI to specify the environment workspace. Of course, the GUI is not available in IDLE, so you will have to use code to specify the environment workspace. I suggest making use of the Python input function to request this information from the user, rather than building the file path directly into the code. After all, you don't know where I keep my files, and I should not have to edit your code at all to grade it. Therefore, it's best if you ask me, as the user, where I have my files. This means using the `input()` function that we have seen before.

Similarly, while it isn't used here, if we were making use of an ArcGISProject object, the "CURRENT" approach to pointing it to the correct project would fail. After all, "CURRENT" is an instruction to use the currently open project in ArcGIS Pro. If you are in IDLE, you probably don't have ArcGIS Pro open to even have a currently open project, so trying to access this, when it doesn't exist, is likely to end in error messages. Therefore, in order to work with and access an ArcGISProject, you'd have to ask the user to specify which .aprx file to use as the ArcGISProject. There may be more changes needed to carry out more complicated ArcGIS processes in IDLE, but these are the only ones needed at this time.

Task 4: Debugging Exercise

The code for this debugging exercise should, when fixed, print out three properties for the first feature class in the environment workspace: the feature type, the OID field name, and the name of the geometry field. I suggest you use the Describe documentation from task 2 to help with tracking down and potential logical errors and the appropriate fixes to the bugs. There are three bugs.

```
desc = arcpy.describe(fcList[0])
print(desc.featureType())
print(desc.OIDFieldName)
print(desc.geometryStorage)
```

Task 5: Non-guided Work: Writing Your Own Script

Now imagine the need to do this for another, unknown, series of datasets. You do not know how many of each there are, so will need to use loops to cycle through the lists.

However, you want to be nice to the user and not report error messages for irrelevant properties. This means you might need to be careful what lines you include in each loop, or use conditional statements to avoid error messages.

Write a script to print output similar to what is below, but only using the properties in question 4, and *only printing the properties which are relevant*.

This script should be able to be run **in IDLE** instead of the Python Window, and ask the user to input the file path for the ArcGIS Project.

Example for one dataset:
Data type: FeatureClass
Name: WV_counties
Shape type: Polygon
Name of field #6: Population
Example for another dataset:
Data type: RasterLayer
Name: LandCover
Band count: 1

Hints

Assume I will use at least one feature class, one raster dataset, and one table. (In other words, no empty lists.)

Assume that, for any relevant dataset, there will be enough fields that field #6 exists.

Use `for` loops to cycle through the lists.

Recall how `print` statements work in Python to ensure there are no errors.

Evaluation

Your code runs properly in IDLE on a different set of datasets from that which is in the common drive, and provides the correct output. What is important is that the output is correct no matter what the inputs are, rather than the specific route that you took to get to the answer.

Rasters

Introductory Comments: GIS-Specific Data Structures

As the discussion of computational data types is inextricably linked with GIS data types, we turn to the two primary data types in GIS: rasters and vectors. Rasters are easier to work with in ArcPy, so that is where we are starting. Raster data structures are conceptually a grid of cells, each of which contains a value (Fig. 9.1).

Once we have the values, the simplest approach to storing this in the computer would be to create a list of values, putting them in order, and making sure the number of rows or columns is also represented so there's a way to turn a one-dimensional list into a two-dimensional grid (Fig. 9.2).

For this simple example, writing out 64 numbers would not be a large burden. However, rasters typically have far more than 64 cells in them. Rasters that are grids of 1000×1000 or more, and therefore have over 1,000,000 cells, are not uncommon. A typical raster, giving elevation in the Morgantown area, is shown below (Fig. 9.3). It has almost 17 million pixels, and with elevation values like 1330.099976, it would take over 56,000 pages to print all the individual values.

As you could therefore predict, this approach can create a very large file size. Conveniently, however, neighboring values often have the same value, such as all the elevation values for the Monongahela River, winding through this raster. There are several data structures which use this similarity of neighboring values to reduce the file size of the rasters. A detailed examination of these is beyond the scope of this book, but you can get a sense of the variety of approaches by returning to the ArcPy Describe documentation from the previous chapter, and take a look at the Raster Dataset properties. You will see a property called `compressionType`, and there are 16 different approaches to reducing the file size supported by ArcGIS Pro version 2.8. Some of the additional resources listed in the Concluding Remarks provide more information about raster data structures.

For the purposes of these introductory remarks, it is sufficient to understand that there are two primary goals that these raster data structures all aim to support: a reduced file size from simply listing every single value and being able to easily identify the value at any given location. The first goal is a straightforward reaction to large file sizes and limited storage abilities. The second is valuable not just for displaying the raster on the screen, but also for the tools that will be examined in the tasks of this chapter. The tools need this information because many raster operations are focal, meaning they use a neighborhood around the central cell, and apply the tool's calculations to all the values within that neighborhood. For example, ArcGIS Pro calculates the slope of the raster at a cell by applying some algebraic calculations of the 3×3 window centered on the cell, not just the central cell.[1] This means that whatever data structure is being used must be able to readily find the neighboring cells to any specified cell. This need to get the values of a neighborhood of cells is enhanced by the ability to find the value at any location.

Conveniently, we will not have to worry about the exact specifications of the compression types as we carry out the remaining tasks. Within ArcPy, most raster processing functions are in the Spatial Analyst, or `.sa` module, so that will be the focus of the following tasks.

---

[1] For all the gory details, go to https://pro.arcgis.com/en/pro-app/latest/tool-reference/3d-analyst/how-slope-works.htm

---

Supplementary Information: The online version contains supplementary material available at [https://doi.org/10.1007/978-3-031-08498-0_9].

J. Conley, *A Geographer's Guide to Computing Fundamentals*, Springer Textbooks in Earth Sciences, Geography and Environment, https://doi.org/10.1007/978-3-031-08498-0_9

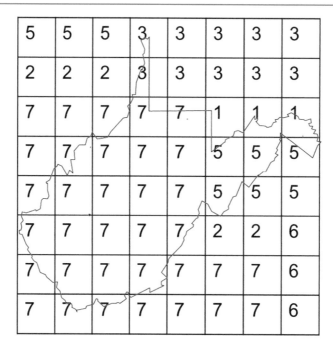

Fig. 9.1 Simplification of a raster

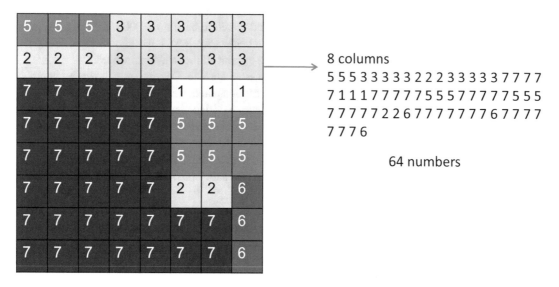

Fig. 9.2 Simplified computational representation of a raster

Task 1: Slope Considerations

Scenario: We will examine rasters through a scenario about red cedar management. You are trying to identify areas in and around Morgantown where red cedar might grow. It prefers steep slopes on south- to southwest-facing aspects. The slope criteria for red cedar are steeper than 30°, ideally steeper than 45°. So we will need to create a slope from the DEM and then reclassify it into these three categories.

Load the morg3m DEM into ArcGIS. When copying it over, make sure to not use any spaces in your file path. Working with rasters is, for whatever reason, one of the areas where ArcGIS is still finicky about spaces in file paths. Before doing anything else, change your environment workspace to go to wherever you saved your data. You may change it through the GUI or through Python. Because the code for the unguided portion of this lab is going to be done one line at a time, we can use the Python Window.

With so much of the raster functionality in the `arcpy.sa` module, we would be typing `arcpy.sa` over and over again throughout the script. Even though ArcPy is already imported into the Python Window, to save ourselves the trouble of typing

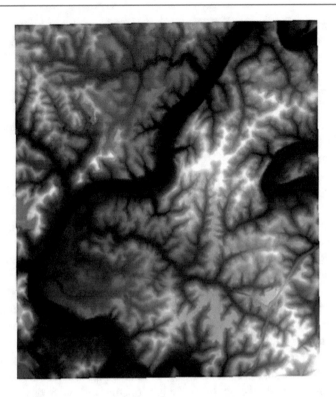

Fig. 9.3 Digital elevation model of Morgantown, WV, area

"arcpy.sa" before every Spatial Analysis command, and thereby cluttering up the script, we can give ourselves direct access using the `from` version of an `import` statement.

```
from arcpy.sa import *
```

Next we will take a look at the ArcGIS help documentation for the Slope command. There is a Slope tool in the 3D Analyst toolbox and the Spatial Analysis toolbox, and they are not identical. We are interested here in the Spatial Analyst version (hence, the .sa module), so the most reliable means of finding the right documentation here is to open up the Arc Toolbox, navigate to Spatial Analyst Tools → Surface → Slope, open that tool, and click the small blue question mark to bring up its documentation.

Question 1: What are the five[2] input parameters of the Slope tool, what do they contain, and which one(s) are required?

Enter the following line to compute the slope:

```
slope = Slope("morg3m")
```

You might get a lengthy error message ending in "Tool Slope is not licensed." If so, you need to enable the spatial analyst extension. Recall that some tools have been put into extensions, which must be enabled before they can be executed. Many of the raster processing tools, being in the spatial analyst toolbox, need to have that toolbox checked out to ensure they get executed.

Enter this line to check it out, and then redo the slope command.

```
arcpy.CheckOutExtension("Spatial")
```

It may still take a while, since Python doesn't speed up raster processing any more than the Slope tool in the toolbox does, but the slope should eventually appear. Once you have this, you should have a variable called `slope` that contains a raster. We will use this to examine some of the properties of rasters.

---

[2]If you have a greater or lesser number of parameters, double-check that you are looking at the documentation for the **slope** tool in the **spatial analyst** package, within **ArcGIS Pro**.

Some of these match what we saw when describing data, such as the band count. As we have done before, if you type in `slope.` and wait, the list of possible methods and functions available to the raster object will appear. One of these will provide the band count, and looking at the list, it should be pretty obvious which one.

Question 2: How many bands are in our slope raster?

Next, let's look at the format for the raster. After all, this could potentially give information about how the raster is really represented in the computer, by telling us its file format.

Type in `slope.format` and press Enter.

Question 3: What happened when you entered `slope.format` and pressed Enter?

This might seem an odd behavior. Recalling all the different data types, perhaps this happened because the format is blank. One way to confirm this is to ask whether the format is equal to `None`.

Type `slope.format == None` and press Enter. Note the double equals sign, to make sure we are testing equality, rather than trying to assign a value. This is important because the format property is read-only, ensuring that we will get an error if we accidentally use a single equals sign and thereby try to assign it a value directly through the property.

You should have been told that this is `True`, confirming that the format is, indeed, nothing. Why would the format be nothing, though? We are able to have rasters which we have not saved to a file. You might have noticed that no output file was included in the parameters in question 1, suggesting that we do not have any output file yet. This is consistent throughout working with rasters, because the raster objects are, as suggested above, routinely very large. Saving and loading such large data structures takes time. For this reason, ArcPy is allowing us to take that time to save the raster only when we actually need it saved for future reference beyond this script. In the list of properties and methods, you may have noticed a `.save()` method, which is used for this exact purpose.[3] We can confirm this status with the following line:

```
slope.isTemporary
```

Question 4: Is the slope a temporary or permanent raster?

We cannot ask the raster to directly tell us which value is in a specific row and column index, probably because it isn't actually represented as a whole series of rows and columns like a spreadsheet. We can, however, ask the raster to tell us the value at a particular location, which is probably more useful anyway. Besides, if we know the corners of the raster, and the cell size, we can use a little algebra to convert back and forth between row/column indices and locations on the ground.[4] This is through a tool that is outside of the spatial analyst toolbox.

Enter `cell = arcpy.management.GetCellValue(slope, "590093 4381728")`[5] and press Enter. The first parameter is the raster in question (`slope` here), and the second parameter is a string containing the X and Y coordinates of the location you want separated only by a space, and these coordinates must be in whatever coordinate system the map is using. (We are using UTM, zone 17N, in this exercise.)

Type `cell` and press Enter.

Question 5: What is the slope at the coordinates (590093, 4281728)?

You should have noticed something slightly unusual about this. Instead of simply giving you the answer, you got a line that starts with the word "Result", as such:

```
>>> cell = arcpy.management.GetCellValue(slope, "590093 4381728")
>>> cell
<Result '[Answer to Q5 redacted]'>
```

---

[3] Perhaps confusing matters some, the slope output should have appeared in the Table of Contents, and would, for that matter, appear in the "scratch geodatabase" that you have through the workspace environment. This is probably the geodatabase tied to the project you should've created for this exercise. It is, still, as you will see shortly, temporary and might not persist after closing and reopening ArcGIS Pro.

[4] It's a little algebra if we assume pixels are aligned perfectly north/south/east/west. If the pixels are at an angle, which is possible, the algebra is still possible but is much more complex.

[5] Older versions of ArcGIS Pro might require arcpy.GetCellValue_management(…).

Confirm this with `type(cell)`. You'll see that this is a special kind of object within `arcpy` called a `Result` object. `Result` objects are what are sometimes called "wrapper" objects, because they "wrap" a bunch of useful information around a theme into a single object. In this case, the theme is the operation of the `GetCellValue` function.

The most useful part for most situations is the output. A result can contain multiple outputs, so we can start with `cell.outputCount` to find out how many outputs there are. It should be pretty straightforward that getting the value of a cell within a raster with only one band will only have one output. Enter `cell.getOutput(0)` to confirm that the output matches your answer to question 5.

Result objects match a critical aspect of how ArcGIS projects retain and store the geoprocessing history. Go to the Catalog pane in ArcGIS Pro, and go to the History tab. You should see a "Get Cell Value" entry there, even though you didn't run it through the toolbox or catalog. This illustrates that running a tool through ArcPy and running it through the GUI are equivalent. In fact, if you right-click on any entry in the History, three of the options relate to the Python version of that entry: "Copy Python Command," "Send to Python Window," and "Save as Python Script." If you hover over the Get Cell Value entry in the History, you'll see details about its operation. It turns out that we can access this information through the `result` object. I'm illustrating below the result of the `slope` operation we did before, so that the following image doesn't give you answers to any questions.

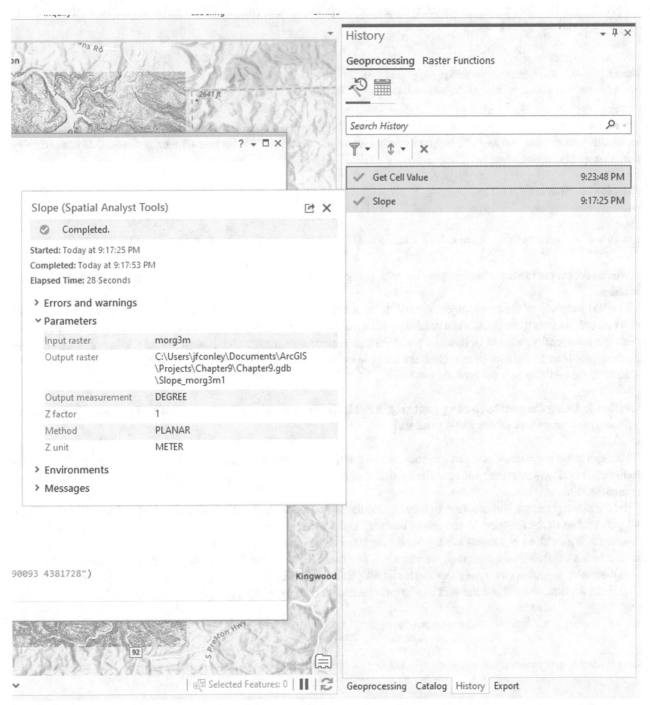

The relevant method names associated with the `result` object are pretty straightforward. We can look at the parameters through the inputs in the `result` object. The slope example I have above shows six inputs, even though I did not specify six inputs in the call to the `slope` function. This is because the `Result` object can still represent the default values within its input list, and this parameter list can include the output, even though the `arcpy.sa.slope` tool doesn't even have an output raster parameter. Likewise, the information that pops up when you hover over the Get Cell Value includes as one of the parameters, the output value.

Question 6: Using `cell.inputCount`, how many inputs were there?

Notice, also, that there are messages. As with counting the inputs and outputs, we can count how many messages there are with `cell.messageCount`. We can also find out what they are with the following command:

```
cell.getMessages()
```

This shows us all the messages at once, as I have below (using a different set of X and Y coordinates, so the answer to question 5 is still not given here).

```
>>> cell.getMessages()
'Start Time: Wednesday, October 6, 2021 9:23:48 PM\n0\nSucceeded at Wednesday, October 6, 2021
9:23:48 PM (Elapsed Time: 0.09 seconds)'
```

It might seem odd that there is one long line here (stretching onto a second line) for three messages. However, buried in the middle, you will see "`\n0\n.`" The `\n` characters represent "new lines."[6] If I print the messages instead of just accessing them, this will become clearer.

```
print(cell.getMessages())
Start Time: Wednesday, October 6, 2021 9:23:48 PM
0
Succeeded at Wednesday, October 6, 2021 9:23:48 PM (Elapsed Time: 0.09 seconds)
```

We can access the messages individually in order, using the `cell.getMessage(index)` method, using 0, 1, or 2 for the index.

The last property of the result object we will look at here is the severity, which represents the type of messages that were created. As you'll see later, when creating your own tools, there are three types of messages: informative messages, which have a severity value of 0, that just provide information; warnings, which have a severity value of 1, that convey potential problems to the user even though the tool did operate and produce an output; and errors, which have a severity value of 2, that tell the user the tool did not finish.

Question 7: Using the `maxSeverity` property, what is the "worst" kind of message this tool produced? How many (if any) warnings and errors did it produce?

Through these commands, you can see how `result` objects are useful in recording what was done and what the script produced. This allows us to save and record information about the outputs and the process, while the script itself only records the inputs to the functions.

Before moving on, we will see another way of calculating the slope. In the documentation for the Spatial Analyst slope tool you looked at for question 1, you might have noticed a recommendation to use a newer tool for the slope: Surface Parameters. It provides a bit more flexibility in how the slope is calculated and "provides a newer implementation." Following the link to the Surface Parameters tool, we can see a few more input parameters allowing us to specify the neighborhood more precisely, including larger neighborhoods and adaptive neighborhoods. To compare against the slope tool, which uses a 3x3 fixed window, we will use the default window specifications here, which are also 3x3 and fixed.

```
slope2 = arcpy.sa.SurfaceParameters("morg3m", "SLOPE")
```

---

[6] Some of you may have encountered this before in strings of file paths, where \n was treated as a new line or \t as a tab character, causing odd errors.

Question 8: Using the `GetCellValue` function, are the value of the slope from the `Slope` tool and the value of the slope from the `SurfaceParameters` tool close to each other (within 0.01 degrees)?

Task 2: Reclassification

Going back to the red cedar scenario, once we have the slope created, using either tool, we need to reclassify the slope to isolate the ideal red cedar conditions. We want to reclassify the slope into three categories:

0 to 30 → 0
30 to 45 → 1
45 to 90 → 2

This requires constructing a remap table, which we can create with the following line. Be careful with the position, type, and number of brackets [] and parentheses ().

```
remap = RemapRange([[0, 30, 0], [30, 45, 1], [45, 90, 2]])
```

The remap table is a list of lists, and each of the three lists has, in turn, three entries inside of it. This setup of nested lists is a pretty common data structure for computer science, because it is a handy way to represent tabular data. We can change the formatting of this remap table list of lists into the following, which more closely mirrors the appearance of a table, and which you'll notice matches the reclassification categories above.

```
[[ 0, 30, 0],
 [30, 45, 1],
 [45, 90, 2]]
```

The first item in each inner list is the starting number for the reclassification range. The second item is the ending number of the reclassification range. The third item is the new value this class is getting. If we had more than three classes, we would have a list of four, or five, or however many lists, but each of these smaller, inner lists would have to have three values matching this order of the lowest value in the range, the highest value in the range, and the new value for the class.

Once the remap table is set up, we can carry out the reclassification itself with this line.

```
slopeReclass = Reclassify(slope, "Value", remap)
```

Question 9: Open the attribute table of the reclassified raster to find out how many cells are in each of the categories. Use this to fill out the table below.

Class ("VALUE" in the attribute table)	Number of cells (aka "COUNT")
0	
1	
2	

Task 3: Aspect Considerations

Using similar scripting to the `Slope` tool or the `SurfaceParameters` tool, create the aspect of this DEM, and reclassify it for the following ranges, because red cedar prefers south- and southwest-facing slopes.

-1 to $135 \rightarrow 0$
135 to $180 \rightarrow 1$
180 to $270 \rightarrow 2$
270 to $360 \rightarrow 0$

Question 10: What is the code you used to create your aspect raster, and the reclassified aspect raster?

Task 4: Map Algebra

Next, to combine the two reclassed maps into a suitability map, we want their product. There are a couple ways of conducting map algebra. The first of these is through the tools within the spatial analyst toolbox. The best means of carrying out this combination step is multiplication. Looking through the different tools available within the spatial analyst toolbox should make clear one that would allow you to multiply two rasters together. For example, if instead of multiplying, we were adding rasters, I could have `suitability = Plus(slopeReclass, aspectReclass)`.

Alternatively, observe that the `.sa` module is the only module which contains its own operators. This means that you can apply operators to add, subtract, multiply, or divide rasters in the same manner that you would add, subtract, multiply, or divide numbers together. The addition version of this approach would give `suitability = slopeReclass + aspectReclass`.

Choose one of those approaches for finding the product of the two reclassed rasters and write your code below.

Question 11: Provide the map algebra code for the product of the two reclassed rasters.

Recall from above that saving map algebra outputs is not automatic. Use the `.save()` method of raster objects to save the suitability raster to file. If you provide a name as the input parameter, it will save the raster as having that name, and located in the environment workspace you specified at the start of task 1.

```
>>> suitability.save("suitability").
```

Task 5: Debugging Exercise

This debugging task repeats the scenario above, but with different criteria, looking for areas which are flatter (slope < 5% is ideal, and < 10% is acceptable) and east-facing (between 45 and 135 degrees acceptable, and between 70 and 110 ideal). It is to calculate suitability by adding the two reclassified inputs, and save it to a new file called "flat_east." There are four bugs.

```
from arcpy.sa import everything
slope_task5 = SurfaceParameters("morg3m", SLOPE)
aspect_task5 = SurfaceParameters("morg3m", "ASPECT")
slope_remap_task5 = RemapRange([[0, 5, 2], [5, 10, 1], [10, 90, 0]])
aspect_remap_task5 = RemapRange([[-1, 45, 0, 45, 70, 1, 70, 110, 2, 110, 135, 1, 135, 360, 0]])
slopeReclass = Reclassify(slope_task5, "Value", slope_remap_task5)
aspectReclass = Reclassify(aspect_task5, "Value", aspect_remap_task5)
suitability_task5 = slope_remap_task5 + aspect_remap_task5
suitability.save("flat_east")
```

Task 6: Unguided Work

Using the same DEM and the Brooks Hall shapefile, write a script to run *in IDLE*, *___not___ the Python Window*, which will complete the following tasks:

1. Ask the user to input the workspace folder where the data is stored.
2. Ensure the spatial analyst extension is checked out.
3. Create a viewshed of areas visible from Brooks Hall.
4. Find areas above 350 meters above sea level.
5. Create a new raster of high-elevation areas which are visible from Brooks Hall.
6. Save that raster as "Lab9Sec5" within the workspace provided in task 1.

Introductory Comments: How a Computer Represents Tables

In Chap. 8, we were introduced to accessing data within ArcPy, and Chap. 9 discussed rasters. This chapter moves toward vector data, but focuses on a superclass of all vector data: tables. Tables are represented within the computer as, in a sense, the grids that the raster data structures worked so hard to avoid. The tables are organized in rows and columns, and each cell corresponding to a row/column combination has exactly one value. The number of entries in the typical table is much fewer than the number of cells in a raster, and neighboring values are much less likely to be the same, so there is no attempt to compress the file size.

To access and manipulate the data in the tables, we use a set of tools called cursors. These cycle through the rows in the table, one at a time, much like the `csv.reader` class allowed us to look at the contents of a csv file. Therefore, they work on tables and any subclasses of tables, namely, vector data, as those are the data types with attribute tables. With the name cursor, you can think of them the same way you think of the cursor in a word processor: something that marks the place where you are. However, instead of marking where in a text document you are typing, these cursors mark the row in a table that ArcGIS is working with. It always follows a few rules, so it isn't as flexible as a cursor in a text document. It must start at the top of the table, just before the first row, so that the first call of the `next()` method, explained below, moves it to the first row. It examines each row one at a time, and each time it finishes with a row, it must move to the next row—no skipping around. Lastly, once it reaches the end of the table, after it has finished examining the last row, the cursor just sits there rather than returning to the top of the table. This last part might seem odd, but becomes useful when we use `for` loops to work with cursors. If it returned to the beginning to keep going, a `for` loop could become an infinite loop and never stop! Figure 10.1 illustrates these rules, with the blue arrows representing the position of the cursor.

Instead of just one type of cursor, ArcPy has provided three different cursors for us to use, all of which are in the `arcpy.da` module (for data access).[1] The first, and simplest, is a search cursor. It is used to access and subset the data, but does not have writing privileges to the dataset. If you want to edit the dataset, you will instead need an update cursor. It allows values in the table to be edited, and entire rows to be deleted. However, it does not add new rows. This requires an insert cursor, whose job is to add new rows to the table.

Task 1: Introduction to Search Cursors

This election feature class has data for many statewide elections, for senate, governor, and president, both primary and general elections. We will use all three kinds of cursors to answer questions and add a couple fields to the feature class, using data from the table.

---

[1]There is also a set of three cursors in the general ArcPy module, but these are emphatically not recommended; they are only provided for "backward compatibility," meaning the ArcPy people don't want to get rid of them just yet because old scripts using then will become broken, but any new scripts should not use them.

Supplementary Information: The online version contains supplementary material available at [https://doi.org/10.1007/978-3-031-08498-0_10].

J. Conley, *A Geographer's Guide to Computing Fundamentals*, Springer Textbooks in Earth Sciences, Geography and Environment, https://doi.org/10.1007/978-3-031-08498-0_10

COOP	Snow	Rain	Max_Temp	Min_Temp
46009404	12.8702	3.464753	35.6261	14.24381
46010205	17.74164	0.004563	32.00648	13.0206
46020204	14.88639	0.793814	33.72843	15.34324
46035505	27.40419	0	28.57812	2.489763
46046206	12.23894	0.565984	34.40068	15.6128
46050904	32.77542	0	23.20795	1.031784
46052704	26.31371	0	28.66026	2.717987
46053104	25.87753	0	31.25737	6.470764
46058004	24.56384	0	26.83511	6.150813
46058204	26.7315	0	25.99305	6.368634
46063302	19.91581	0.83035	33.14756	8.761695
46066704	26.23352	0	26.98605	1.961624
46071206	10.17711	4.251706	37.5514	19.66586
46078402	10.361	3.872977	36.77487	18.47435
46084204	15.70447	1.0544	33.9156	13.06227
46092105	30.7403	0	28.97272	1.185583
46092505	29.79927	0	25.52902	3.30918
46093905	15.37704	0.292144	33.79165	16.67643
46107503	8.544963	0.499819	36.76236	20.84514
46121504	23.08866	0	31.52944	5.780627
46121802	15.80627	2.001364	34.53439	16.05507
46122002	17.02071	1.110644	34.92959	16.47385
46128202	8.646267	3.101837	38.43576	22.62863
46132306	11.12392	1.467303	34.16127	19.91913
46132406	11.81825	3.160169	36.6137	19.46284
46133001	10.01655	0.201227	36.32031	21.50096
46136304	24.66696	0	29.63738	4.392129
46139704	33.94301	0	23.06192	-1.23523
46157003	11.01118	1.650091	34.89165	21.27322
46157803	7.53162	2.974514	37.96117	21.11567
46157903	11.32947	4.581947	37.24768	18.63561
46158003	6.321599	1.818262	35.7416	22.27161
46158806	6.938579	6.782516	39.11125	24.76188
46167702	10.94839	0.70966	35.0351	19.92853
46169603	8.764785	2.860696	35.40187	23.61442
46172303	8.34917	2.016127	38.41698	22.80826
46190004	24.59475	0	29.746	6.810268
46205401	7.73484	1.625156	34.96359	20.65997
46215104	21.72153	0	28.97286	8.559306
46215601	14.45292	0.16171	34.66978	13.02769
46246204	13.81076	0.13822	32.40534	14.94112
46252203	13.49281	0.133752	32.99892	16.19504
46262203	8.191582	0.850091	36.10131	23.56093
46269701	8.09697	0.428481	36.26493	21.0765
46270905	20.33721	0	31.23017	12.11832
46271604	21.46613	0	31.95239	7.4091

COOP	Snow	Rain	Max_Temp	Min_Temp
46009404	12.8702	3.464753	35.6261	14.24381
46010205	17.74164	0.004563	32.00648	13.0206
46020204	14.88639	0.793814	33.72843	15.34324
46035505	27.40419	0	28.57812	2.489763
46046206	12.23894	0.565984	34.40068	15.6128
46050904	32.77542	0	23.20795	1.031784
46052704	26.31371	0	28.66026	2.717987
46053104	25.87753	0	31.25737	6.470764
46058004	24.56384	0	26.83511	6.150813
46058204	26.7315	0	25.99305	6.368634
46063302	19.91581	0.83035	33.14756	8.761695
46066704	26.23352	0	26.98605	1.961624
46071206	10.17711	4.251706	37.5514	19.66586
46078402	10.361	3.872977	36.77487	18.47435
46084204	15.70447	1.0544	33.9156	13.06227
46092105	30.7403	0	28.97272	1.185583
46092505	29.79927	0	25.52902	3.30918
46093905	15.37704	0.292144	33.79165	16.67643
46107503	8.544963	0.499819	36.76236	20.84514
46121504	23.08866	0	31.52944	5.780627
46121802	15.80627	2.001364	34.53439	16.05507
46122002	17.02071	1.110644	34.92959	16.47385
46128202	8.646267	3.101837	38.43576	22.62863
46132306	11.12392	1.467303	34.16127	19.91913
46132406	11.81825	3.160169	36.6137	19.46284
46133001	10.01655	0.201227	36.32031	21.50096
46136304	24.66696	0	29.63738	4.392129
46139704	33.94301	0	23.06192	-1.23523
46157003	11.01118	1.650091	34.89165	21.27322
46157803	7.53162	2.974514	37.96117	21.11567
46157903	11.32947	4.581947	37.24768	18.63561
46158003	6.321599	1.818262	35.7416	22.27161
46158806	6.938579	6.782516	39.11125	24.76188
46167702	10.94839	0.70966	35.0351	19.92853
46169603	8.764785	2.860696	35.40187	23.61442
46172303	8.34917	2.016127	38.41698	22.80826
46190004	24.59475	0	29.746	6.810268
46205401	7.73484	1.625156	34.96359	20.65997
46215104	21.72153	0	28.97286	8.559306
46215601	14.45292	0.16171	34.66978	13.02769
46246204	13.81076	0.13822	32.40534	14.94112
46252203	13.49281	0.133752	32.99892	16.19504
46262203	8.191582	0.850091	36.10131	23.56093
46269701	8.09697	0.428481	36.26493	21.0765
46270905	20.33721	0	31.23017	12.11832
46271604	21.46613	0	31.95239	7.4091

COOP	Snow	Rain	Max_Temp	Min_Temp
46009404	12.8702	3.464753	35.6261	14.24381
46010205	17.74164	0.004563	32.00648	13.0206
46020204	14.88639	0.793814	33.72843	15.34324
46035505	27.40419	0	28.57812	2.489763
46046206	12.23894	0.565984	34.40068	15.6128
46050904	32.77542	0	23.20795	1.031784
46052704	26.31371	0	28.66026	2.717987
46053104	25.87753	0	31.25737	6.470764
46058004	24.56384	0	26.83511	6.150813
46058204	26.7315	0	25.99305	6.368634
46063302	19.91581	0.83035	33.14756	8.761695
46066704	26.23352	0	26.98605	1.961624
46071206	10.17711	4.251706	37.5514	19.66586
46078402	10.361	3.872977	36.77487	18.47435
46084204	15.70447	1.0544	33.9156	13.06227
46092105	30.7403	0	28.97272	1.185583
46092505	29.79927	0	25.52902	3.30918
46093905	15.37704	0.292144	33.79165	16.67643
46107503	8.544963	0.499819	36.76236	20.84514
46121504	23.08866	0	31.52944	5.780627
46121802	15.80627	2.001364	34.53439	16.05507
46122002	17.02071	1.110644	34.92959	16.47385
46128202	8.646267	3.101837	38.43576	22.62863
46132306	11.12392	1.467303	34.16127	19.91913
46132406	11.81825	3.160169	36.6137	19.46284
46133001	10.01655	0.201227	36.32031	21.50096
46136304	24.66696	0	29.63738	4.392129
46139704	33.94301	0	23.06192	-1.23523
46157003	11.01118	1.650091	34.89165	21.27322
46157803	7.53162	2.974514	37.96117	21.11567
46157903	11.32947	4.581947	37.24768	18.63561
46158003	6.321599	1.818262	35.7416	22.27161
46158806	6.938579	6.782516	39.11125	24.76188
46167702	10.94839	0.70966	35.0351	19.92853
46169603	8.764785	2.860696	35.40187	23.61442
46172303	8.34917	2.016127	38.41698	22.80826
46190004	24.59475	0	29.746	6.810268
46205401	7.73484	1.625156	34.96359	20.65997
46215104	21.72153	0	28.97286	8.559306
46215601	14.45292	0.16171	34.66978	13.02769
46246204	13.81076	0.13822	32.40534	14.94112
46252203	13.49281	0.133752	32.99892	16.19504
46262203	8.191582	0.850091	36.10131	23.56093
46269701	8.09697	0.428481	36.26493	21.0765
46270905	20.33721	0	31.23017	12.11832
46271604	21.46613	0	31.95239	7.4091

Fig. 10.1 Cursor positions

Item 1: In the 2012 Democratic primary, President Obama lost several counties to political unknown Keith Judd.[2] We will use a search cursor to identify which counties these are. As the purpose of a search cursor is to subset the data, it usually uses a query. This is not required, however, if you want to read (but not edit) data from all rows in the table. The first line below creates the cursor, and it is very similar for all three types of cursors, at least in the terms of which parameters are required or frequently used. These are the three parameters used below, which are the first three that are described in the documentation. Please use the documentation to answer question 1.

Question 1: The first three parameters for the arcpy.da.SearchCursor() function are the ones you are most likely to use. What is the role of each of them?

Open a Python Notebook and run the following lines, being careful about the variable names. The output of this is the answer to question 2, so it is not printed below.

```
search = arcpy.da.SearchCursor "WV_elections2016", ["Name"], '"Obama12PPV" < "Judd12PPV"')
for county in search:
 print(county[0])
```

Question 2: (10 points) Which counties gave more votes to Mr. Judd than Mr. Obama?

You may notice that the `print` statement accesses the name through `county[0]`. The use of brackets is a clue about `county`'s nature: it is a list. Not only that, but it follows a predictable structure based upon the code you give it. Recall from this chapter's introductory remarks that the cursor must proceed through the table one row at a time. Each time through the loop, `county` is referring to the next row in the Table. At first, you might expect then that it is a list of all the values of the cells in that row. However, that would be more cells than we need. Instead, each item in the list contained in the `county` variable is the value corresponding to the same index in the list of attributes from the second input parameter. This means that in the example above, `county` is a list of exactly one item: the name of the county. With a query specified, the `for` loop will only access

---

[2]Also, inmate at a federal penitentiary.

entries which meet that cursor's criterion. This means that the SearchCursor represented by the `search` variable has selected out only those counties which voted for Judd in this primary election. It is as if the search cursor checks the first row for whether that query's criteria are met, and provides that row if and only if the criteria are met. If not, it skips to the next row and checks that row, and continues the process until it either reaches the end of the table or it reaches another row that meets the criteria and provides that row. In this way, only the subset of counties which voted for Judd is selected out without having to use the `Select()` function as in Chap. 7.

Task 2: Introduction to Update Cursors

Item 2: In the Republican primary of that year, the number of total votes seems to be mistakenly calculated, since all the values are zero. We will want an update cursor to fix this problem. Because this now requires editing the data, we need to have an update cursor, rather than a search cursor.

First, let's create the update cursor. Enter the following line into a new Notebook cell. You will notice its similarity to how the search cursor was created. It is, indeed, the same syntax and same input parameters, but without the search query this time. Again, pay close attention to variable names: there are many similar variable names, and all the candidate names are all truncated to five characters to fit a naming convention.[3]

```
update = arcpy.da.UpdateCursor("WV_elections2016", ["Romne12PPV", "Santo12PPV", "Paul12PPV",
"Gingr12PPV", "Roeme12PPV", "Tot12Rep"])
```

Now, to make the corrections, we will calculate the total votes cast field as the sum of the individual candidates' totals, and add those to the table. As with the search cursor, `county` is a list referring to each row's values as the cursor cycles through the rows one by one, and the elements of that list are the values corresponding to the list in the second input parameter to the `UpdateCursor` constructor method that created the `update` variable.

```
for county in update:
 county[5] = county[0] + county[1] + county[2] + county[3] + county[4]
 update.updateRow(county)
```

When trying to edit files, you may be familiar with messages that tell you a certain file is locked for editing by another application. Update cursors like this are one of the applications that can and will do this. This means we have to remove that lock on editing to make sure things behave well if we want to use another update or insert cursor. This is accomplished by deleting the variable storing the update cursor, because deleting it also deletes any associated locks from memory. This uses a new keyword: `del`.

```
del update
```

Look in the attribute table. The Tot12Rep column should be filled in with numbers now. Right-click on the column header and choose statistics.

Question 3: What are the mean and standard deviation of the total votes cast by county in the 2012 Republican presidential primary?

Task 3: Deleting and Inserting Rows

Imagine you have just received the results from the 2018 election, and want to add it into your feature class of election results so that you may compare Senator Manchin's results in 2018 against those from 2012. However, this is not a straightforward join, as the structure of the 2018 election results doesn't completely match that of our general feature class. Open up the table

---

[3]Candidate (5 characters) + year (2 characters) + office (**P**resident, **G**overnor, or **S**enate) + election type (**P**rimary or **G**eneral) + attribute type (**V**ote total or **P**ercent).

in the .gdb that was provided. You should notice a few things: First, the second row is a description of the headers. While this is useful information, it does not have a counterpart in the feature class, so we will want to remove it. Also, if you look closely, you might notice that a county is missing: Monongalia County! So we want to delete the information in the second row, and insert a new row for Monongalia County.

To delete a row, we again want an update cursor, but this time, for the table with the 2018 results. First, notice that we are only accessing the name attribute. Second, there is a query to ensure we are only looking at the information row, by asking whether the attribute "Name" matches the value "County Name." You might notice that the attribute name (of NAME) is in double quotes, while the value of the attribute (County Name) is in single quotes. This is how SQL is set up. However, to achieve the string of "NAME" = 'County Name', we need to use the escape character \' to indicate that we want to include a single quote mark at this spot in the string. It is needed to allow Python to recognize the difference between a quote mark that is a character in the text string and a quote mark that ends the string.

Lastly, because deleting rows is not something that can be undone, I have written a double check into the code, by again seeing if the name has the value of "County Name." This way, we are making absolutely sure we do not accidentally and permanently delete the wrong row(s). At the end, we will again need to delete the cursor.

```
update = arcpy.da.UpdateCursor('WV_election_2018_only', ["NAME"], '"NAME" = \'County Name\'')
for row in update:
 print(row)
 if row[0] == "County Name":
    update.deleteRow()
del update
```

Next, we need an insert cursor to add the data for Monongalia County. The insert cursor gets set up similar to the other two, with the name of a layer, and a set of attributes. This time, however, we are going to tell the computer what values to put into these attributes for the new row(s) instead of extracting those cell values from the table. As you enter the code below, take special care with the tuple of the second line. Note that there are two pairs of parentheses. The first indicates the input parameter for the insertRow() function. The input parameter is itself a tuple, also using a pair of parentheses, providing the values that will be added to the table. In this situation, be careful and read over your values (at least) twice before running the cell in the Notebook. These values must be in the same order as the list of parameters used when creating an insert cursor, because the computer is not smart enough to rearrange them if you made any mistakes. It is a challenge (but possible) to edit them after the fact.

```
insert = arcpy.da.InsertCursor('WV_election_2018_only', ["NAME", "FIPS", "Manch18SGP", "Morri18SGP",
"Holle18SGP"])

insert.insertRow(("Monongalia", 54061, 58.19627104, 35.2796717, 6.524057259))

del insert
```

After running these lines, close the table (if it was open) and reopen it. Take a look to confirm that the insert cursor worked.

Question 4: What is the value for Monongalia County for most attributes? Why did this happen?

Task 4: More Advanced Use of Cursors

If this has been done correctly, we can now use cursors again to put the 2018 senate general election percents into the existing feature class. We only want these columns, not the entire set of 2018 attributes, because with the two elections for the state supreme court, and the amendments, there are many results in this 2018 election file that we do not need. This means that a join might be overkill, even though the data is there and (if task 3 was done correctly) done in such a way that a join is now possible. Instead, we will use cursors to add just the three desired columns.

Before this happens, though, it is helpful to consider what is needed, and plan out the task with a flow chart.

The core of the task will be taking the values that are extracted from the table and put them into the feature class. It should by now be clear this will require an update cursor. So let's start out our flow chart.

This will need some fleshing out, of course. The first expansion to this is that we need to figure out how we are getting the 2018 results into the existing feature class. To do this, we need a place to put them—in other words, new attributes, one for the percent of the vote total won by each of the three candidates: Joe Manchin (Democrat), Patrick Morrisey (Republican), and Rusty Hollen (Libertarian). It should be clear that we want to do this before we can run the for loop, so let's add it to the start of the flow chart.

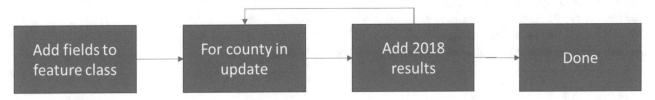

This means we will have a place to put the data from the stand-alone table, but doesn't tell us how we get the data out of the table. A first thought you might have is to have two cursors running in parallel in the same for loop, as below. However, this is not likely to work, and even if it does, it is not guaranteed to work every time.

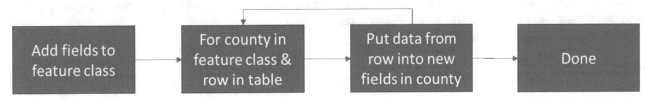

Question 5: Why would this not be guaranteed to work? (If you're stumped, take a look at both the attribute table for the feature class and the table, and consider stepping through row by row in each.)

For the reason I hope you came up with in question 5, we will instead want to use a dictionary for each of the three candidates, matching each county's FIPS code with the vote percent for the relevant candidate. Populating these dictionaries will take a separate for loop using a search cursor to access each row in the Table.

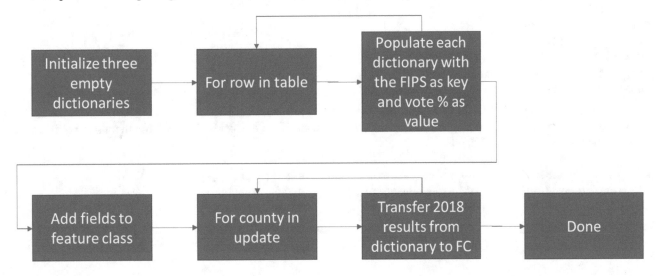

Now that our flow chart is done, we can turn that into pseudocode, as below:

ManchinDict ← new dictionary.
MorrisseyDict ← new dictionary.
HollenDict ← new dictionary.
Search ← SearchCursor for 2018 results, needing FIPS, Manchin %, Morrissey %, and Hollen %.

For each row in Search:

→ ManchinDict—associate FIPS with Manchin %.
→ MorrisseyDict—associate FIPS with Morrissey %.
→ HollenDict—associate FIPS with Hollen %.
Add Manchin 2018% field to feature class.
Add Morrissey 2018% field to feature class.
Add Hollen 2018% field to feature class.
Update ← Update Cursor for feature class, needing FIPS and newly created fields.

For each county in Update:

→ Manchin 2018% ← the value in ManchinDict associated with the key FIPS
→ Morrissey 2018% ← the value in MorrisseyDict associated with the key FIPS
→ Hollen 2018% ← the value in HollenDict associated with the key FIPS

With this pseudocode created, we can now turn it into the python code below, which I am presenting in stages. You may enter this in stages into the Python Notebook, or all at once.

The new dictionaries are created first.

```
ManchinDict = {}
MorrisseyDict = {}
HollenDict = {}
```

Next, we create the search cursor for the 2018 results. Note that this will only work if you already have the table loaded into ArcGIS Pro.

```
search = arcpy.da.SearchCursor('WV_election_2018_only', ["FIPS", "Manch18SGP", "Morri18SGP",
"Holle18SGP"])
```

Now we can run the loop through the search cursor.

```
for county in search:
 fips = county[0]
 ManchinDict[fips] = county[1]
 MorrisseyDict[fips] = county[2]
 HollenDict[fips] = county[3]
```

With this loop done, and the dictionaries populated, we can then create our fields. This is through an `arcpy` function called `management.AddField()`.

```
arcpy.management.AddField('WV_elections2016', "Manch18SGP", 'DOUBLE')
arcpy.management.AddField('WV_elections2016', "Morri18SGP", 'DOUBLE')
arcpy.management.AddField('WV_elections2016', "Holle18SGP", 'DOUBLE')
```

Question 6: One of the input parameters has the value "DOUBLE." What does "DOUBLE" mean here for adding a field?

Next, we need to create an update cursor to allow these new fields to get populated with values.

```
update   =   arcpy.da.UpdateCursor('WV_elections2016',   ["FIPS",   "Manch18SGP",   "Morri18SGP",
"Holle18SGP"])
```

Once we have our cursor, we can use a for loop to cycle through it.

```
for row in update:
 fips = str(row[0])
 row[1] = ManchinDict[fips]
 row[2] = MorrisseyDict[fips]
 row[3] = HollenDict[fips]
 update.updateRow(row)
```

Lastly, we should make sure things will continue to run properly by deleting our cursors and any and all associated locks they have placed on the files.

```
del search
del update
```

Using the same approach to answering question 3, answer the following question for the new variables you just created.

Question 7: What are the mean and standard deviation of the percent of votes cast for each of the three candidates in the 2018 senate general election?

Task 5: Debugging Exercise

This script, when fixed, will use a search cursor to identify the counties won by Senator Joe Manchin in 2018, and calculate the average population of those counties. As always, be careful about logical errors that allow the script to finish, but not provide the correct answer. There are three bugs.

```
total_pop = 0
total_count = 0
search = arcpy.da.SearchCursor('WV_elections2016', '"Manch18SGP" > "Morri18SGP"', ["Population"])
for county in search:
 total_pop += county[1]
 total_count += 1
Manchin_pop_mean = total_pop - total_count
delete search
```

Task 6: Writing Your Own Script

Write a script to accomplish the following tasks. Here are your specifications:

1. Add a field, called "Manch1218d," as a double data type.
2. Give it the value of the difference between the 2012 vote percent (Manch12SGP) and the 2018 vote total (Manch18SGP). Calculate it as the *2012 percent minus the 2018 percent*.

3. Print the names of those counties where Senator Joe Manchin's vote percent *increased* from the 2012 general election (Manch12SGP) to the 2018 general election (Manch18SGP).

Hints

Assume I can set the workspace environment through the GUI, so do not include that in your script. Likewise, assume I have any and all data already loaded into ArcGIS Pro, so you do not need to load them through the script.

Make certain you use **SGP** variables in all calculations.

Evaluation

I will run your code in the Python Notebook and see if the output is correct. If the script does not run at all, or gives erroneous output, I will give partial credit based on how many changes are needed to correct it.

Introductory Comments: Data Structures for Points, Lines, and Polygons

In the previous chapter, we used cursors to access the attribute table of a feature class. In this chapter, we will examine and work with the geometry portion of a feature class. Recall that these are vector data, which means our geometries are stored as points, lines, and polygons.

As far as the computer is concerned, the basic building blocks of all these shape types are points, with X and Y coordinates, recorded either in latitude and longitude, or in a planar coordinate system such as UTM or State Plane Coordinate System. Most commonly, these points will be labeled with some name or identifier, and given X and Y coordinates, as illustrated in the table shown below.[1] It should be noted here that what follows is but one approach to representing vector geometries in a computational structure. Different vector file types (shapefile, geodatabase, KML, etc.) can have different underlying structures. The common theme is the use of points, lines, and polygons and ensuring that lines and polygons are comprised of individual points that represent the vertices of lines and polygons (Fig. 11.1).

If our data type is points, we can stop at that. If, instead we have lines, we have another step. This step, though, is as easy as connecting the dots. Lines are defined as a connected set of points. This can be represented through an additional table, as shown below (Fig. 11.2).

Polygons add one additional level, representing the boundary of the polygon as a series of lines which collectively start and finish at the same location, as illustrated below. Often, these lines represent shared boundaries between polygons. For example, we could represent West Virginia as the border between West Virginia and Maryland, the border between West Virginia and Virginia, the border between West Virginia and Kentucky, the border between West Virginia and Ohio, and the border between West Virginia and Pennsylvania. Doing this can then allow us to reuse lines, such that the border between West Virginia and Ohio can be used as part of the outline of the West Virginia polygon and as part of the outline of the Ohio polygon (Fig. 11.3).

Carrying out spatial analysis tasks, from identifying whether a point is in a polygon to carrying out a geographically weighted regression, is all built upon operations using geometry tables like these. After all, a computer can only work with numbers, and it is through tables like these that the computer transforms a geometric shape into a series of numbers. The operations for working with these tables can get surprisingly complex, so a detailed examination will not be undertaken here. If interested, some of the more advanced references in the Concluding Remarks provide examples of how complex it can get, even for something that would seem as simple as identifying whether a point is in a polygon or whether two lines intersect. In this chapter, we will instead use ArcPy to examine, edit, and do some simple geoprocessing functions on vector geometries.

---

[1]As we will see later, the reality of linking coordinates to an attribute table is a bit more complex, but this will work well for illustrating concepts.

Supplementary Information: The online version contains supplementary material available at [https://doi.org/10.1007/978-3-031-08498-0_11].

J. Conley, *A Geographer's Guide to Computing Fundamentals*, Springer Textbooks in Earth Sciences,
Geography and Environment, https://doi.org/10.1007/978-3-031-08498-0_11

Point	X	Y
Pt3	45676	37450
Pt4	45769	37053
Pt5	46289	37054
Pt6	46301	36398
Pt7	46785	36433
Pt8	47014	35879

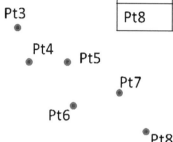

Fig. 11.1 Point geometries and an associated table

Point	X	Y
Pt3	45676	37450
Pt4	45769	37053
Pt5	46289	37054
Pt6	46301	36398
Pt7	46785	36433
Pt8	47014	35879

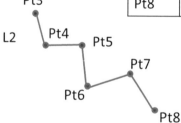

Line	Points					
L2	Pt3	Pt4	Pt5	Pt6	Pt7	Pt8

Fig. 11.2 Line geometry and associated tables

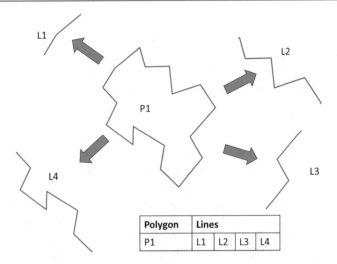

Fig. 11.3 Polygon geometry and an associated table

Task 1: Access Shape Information

Copy the data to your workspace and add all data to a new map. We will be working with the Python Notebook for the guided tasks here, although all of it can be done within the Python Window if your version of ArcGIS Pro does not support Notebooks.

We will now use the shape information encoded in ArcGIS and accessed through ArcPy to examine some properties of each of the three primary types of vector shapes. Recall from previous chapters that vector feature classes are all subclasses of tables. Since we accessed the items in a table through one of three forms of cursors and subclasses inherit functionality from their superclasses, we can also use any of the three kinds of cursors on our feature classes, too. This works because, while in the computer, they are stored separately (e.g., .dbf versus .shp for shapefiles); in practice, the geometry and the attributes are stored together, such that the file for the table and the file for the geometry must be in the same folder, alongside the .shx index file with the information on linking the geometry with the table. In fact, the geometry of a feature class is hidden in plain sight within the attribute table. Notice that, in any attribute table for explicitly spatial data, there will be a column called "Shape," often the second column from the left, although this isn't guaranteed. All of this column's contents in the attribute table will tell you the type of geometry it is (e.g., "Point"). Embedded within that geometric shape, although not readily accessible through the attribute table, is the geometry information of that feature. Even though the attribute table doesn't give us direct access to the geometries, we can use the same ArcPy cursor to access it. We access the geometric information by a special built-in variable name, "SHAPE@," into the list of variables to be accessed when constructing the cursor with `SearchCursor()`, `UpdateCursor()`, or `InsertCursor()`.

Points, lines, and polygons are all accessed through a `Geometry` object, which means that all properties are available for all shape types, even if there isn't always a reason to use it, like getting the area of a point geometry. If you know the specific piece(s) of information within the geometry that you need, there are properties that you can add after the @, such as getting the area through SHAPE@AREA. These are in the Geometric Tokens segment of the help documentation at https://pro.arcgis.com/en/pro-app/arcpy/get-started/reading-geometries.htm, but we will use the generic geometry SHAPE@ for this chapter. It allows you to access all information, although not always as directly.

Let's start with the points, which we have in a layer called "Office_cities." To access the geometries, run the following line:

```
cityCursor = arcpy.da.SearchCursor("Office_cities", ["SHAPE@", "DESCRIPT"])
```

Now we will want to examine how geometries work by getting a single geometry out of this cursor. In the previous lab, we saw how to use a for loop to cycle through the entries of a table. Here, to explore a single geometry, we just want a feature from that cursor. Just as we used the `.next()` function to examine a CSV file one line at a time, we can use the cursor's `next()` function to go through the cursor's contents one line at a time, or as here, to limit us to getting the first item. Run the following line:

```
city = cityCursor.next()
```

The behavior of a cursor does not change just because we are interested in the shape now, so the variable city contains a tuple of items, given in the order in which they are listed in the line defining the cursor. This means that the geometry is in entry #0 within the list. Therefore, access that geometry by running the following line:

```
geom = city[0]
```

We can confirm that this is, indeed, a geometry by typing `geom` into a blank Notebook cell and running it. You will then see the point that this geometry represents. Of course, a point isn't very interesting here, but this might be of more use when we have a more complex shape. Regardless of the type of the geometry, some of the properties of any geometry we may want to look at are the following: how many points it contains, the first and last points, its length, its area, and how many parts it has. The last of the following lines provides something called the "WKT" or "Well Known Text," which is a text-based representation of the geometry. Its advantage is that it is portable; many other systems can import WKT (why it is well-known!), and this would then be a means of taking geometries out of ArcGIS and into other systems.

Using the following set of lines, get the information to answer the subsequent questions. While not all are needed to answer these questions, please run the lines anyway, as they are useful for answering later questions involving lines or polygons, and to illustrate how all properties are given for all geometry types, even if things like the length or the area of a point is a bit uninteresting:

```
print("Point Count: ", geom.pointCount)
print("First Point: ", geom.firstPoint)
print("Last Point: ", geom.lastPoint)
print("Length: ", geom.length)
print("Area: ", geom.area)
print("Part Count: ", geom.partCount)
print("Well Known Text: ", geom.WKT)
```

Question 1: What is the area listed for this geometry?
Question 2: How many parts does this geometry have?
You may have noticed that while the WKT provides text that can be parsed for extracting the coordinate information, it isn't easily done, and the first and last points give all of the coordinates (including two that are not used here) through a Point object, rather than direct numbers. You can confirm this by running `type(geom.firstPoint)`. We have to use this point object, although to show how we can access any point in a geometry that contains many points, not just the first and last points, we will use a different approach, with a function through the geometry object: `getPart()`. Run the next four lines in a Notebook cell to create a point from the first (and only) point in this geometry, and print off its x and y coordinates. Within some rounding tolerance, these should match the coordinates in the WKT:

```
cityPoint = geom.getPart()
x = cityPoint.X
y = cityPoint.Y
print("X: ", x, ", Y: ", y)
```

```
X: -80.087076 , Y: 38.220057
```

Next, we will look at line data. There is a layer called "Trails_in_WMA." Create its cursor and access the first trail's geometry with the following lines:

```
trailCursor = arcpy.da.SearchCursor("Trails_in_WMA", ["SHAPE@", "TRAIL_NAME"])
trail = trailCursor.next()
geom = trail[0]
```

Having used the same variable name for the geometry, `geom`, we can simply run the same Notebook cell as before to print off the geometry's information, like the point count, first point, length, and area. Rerun that cell and use the printed output to answer the questions below:

Question 3: How many points are in this geometry?

Question 4: How many parts does this geometry have?

Question 5: Are the first and last points the same?

We will now access the first part of the geometry to illustrate the geometry object for lines:

```
trailLine = geom.getPart(0)
trailLine
```

Upon entering the second line, you will notice that the line is stored as a type of Python collection we have not yet encountered: the array. Arrays are a lot like lists, but whereas lists can contain any multitude of data types, arrays are constrained such that they can only contain a single data type. Every item in this geometry is a point, meaning this is an array of points. As an array of points, we can think of the line as what would happen playing "connect the dots" with these points in the order in which they are presented. Like lists, we can loop through arrays with a `for` loop or access individual items using an index in brackets.

Question 6: Rounding to the nearest meter, which is the unit for this geometry, what are the X and Y coordinates for the point with index #5?

To illustrate the type constraint, try appending the number 42 to the array with the following line. Upon attempting to run it, you should get an error message with the key information at the end: that you are trying to add an item which is neither a point nor an array that contains only points:

```
trailLine.append(42)
-------------------------------------------------------------------
ValueError                              Traceback (most recent call last)
In [18]:
Line 1: trailLine.append(42)

File C:\Program Files\ArcGIS\Pro\Resources\ArcPy\arcpy\arcobjects\mixins.py, in append:
Line 188: self.add(value)

File C:\Program Files\ArcGIS\Pro\Resources\ArcPy\arcpy\arcobjects\arcobjects.py, in add:
Line 96: return convertArcObjectToPythonObject(self._arc_object.Add(*gp_fixargs(args)))

ValueError: Array: Add input not point nor array object
-------------------------------------------------------------------
```

Lastly, to do the same for polygon data, we will use one of the wildlife management areas. Everything should look very similar to the line, except the list of points in the polygon closes the loop to distinguish it from a line, and the header for the WKT indicates it is a multipolygon, which means that this feature could have more than one polygon in it, like an archipelago, even though it only has one part:

```
wmaCursor = arcpy.da.SearchCursor("WMA_0704_utm83", ["SHAPE@", "NAME"])
wma = wmaCursor.next()
geom = wma[0]
```

Again, by access the same information using the same cell as above, answer the questions below:

```
geom.pointCount
geom.firstPoint
geom.lastPoint
geom.length
geom.area
geom.partCount
geom.WKT
```

Question 7: What is the area of this geometry?
Question 8: Are the first and last points the same?

As with the line, we will access the first (and only) part of this geometry through the lines below. As you run these lines to see the contents of the polygon geometry, you should notice that it is set up *very* similarly to the line geometry. In fact, both are arrays of points, with the only difference being that for a polygon, the first and last points are the same, whereas they are different in a line:

```
wmaPolygon = geom.getPart(0)
wmaPolygon
```

To illustrate another means by which lists and arrays are similar, we can find out how many items there are in the array through the following line:

```
len(wmaPolygon)
```

Based upon the length of the array, as well as the point count property from the geometry object, you should see that it reports seven points. However, if you examine the shape itself by typing `geom` into a cell and looking at it, you should see that it is an irregular hexagon, meaning the polygon has six sides and six vertices.

Question 9: Why does an irregular hexagon have a point count and array length of 7 points?

Task 2: Use Geometry-Related Methods

Scenario: trail management. You have a series of trails in wildlife management areas and offices to manage them. The offices are located in five towns, which are pulled out in the Office_cities file.

Using the geometry help documentation (http://pro.arcgis.com/en/pro-app/arcpy/classes/geometry.htm), identify the method that best gives the distance between two geometries. Recognizing `city[0]` is the geometry of the city and `trail[0]` is the geometry of the trail in the following code, write the expression that evaluates to the distance from a city to a trail.

Question 10: Write an expression for the distance from a single city to a single trail.

Then, try the following code, substituting your answer to Question 1 in the appropriate location, which should give the distances below, with respect to Marlinton:

```
trailCursor = arcpy.da.SearchCursor("Trails_in_WMA", ["SHAPE@", "TRAIL_NAME"])
cityCursor = arcpy.da.SearchCursor("Office_cities", ["SHAPE@", "DESCRIPT"])
for city in cityCursor:
→ for trail in trailCursor:
→ → print ("The distance from ", city[1], " to ", trail[1], " is ", str(***Answer to Question
10***))
```

```
The distance from Marlinton town to Durbin to Cass Connector Rail-Trail is 0.3630134655588971
The distance from Marlinton town to Blackwater Canyon Rail-Trail is 0.9699017185940851
The distance from Marlinton town to Shaver's Fork Valley Trail is 0.3357517019854554
The distance from Marlinton town to Long Point Trail is 0.7747266850184706
The distance from Marlinton town to Shavers Fork Road Trail is 0.772048417334362
The distance from Marlinton town to Allegheny Trail is 1.0520851710706902
```

These distances might all seem oddly low (as in near or below 1.0). This is because the answers could be in decimal degrees. We have data in different projections, and while ArcGIS Pro and ArcPy are able to convert from one to another, they don't always convert in the direction we would prefer. In this case, one file (Office_cities) is in decimal degrees, and the other file (trails) is in UTM coordinates. Instead of risking the chance that ArcGIS will continue to convert everything to the decimal degrees from the Office_cities file, we should reproject Office_cities to use the UTM coordinate system of the other data. The following lines should resolve that, of course, replacing "N:/yourLocation" with the appropriate file path. Note the use of the .prj file, which stores the projection information for any shapefile that happens to have a .prj file attached:

```
inputFeatures = "N:/yourLocation/Office_cities.shp"
outputFeatures = "N:/yourLocation/Office_cities_proj.shp"
projection = "N:/yourLocation/Trails_in_WMA.prj"
arcpy.management.Project(inputFeatures, outputFeatures, projection)
```

This does two things. First and foremost, it creates a new file, Office_cities_proj.shp, that has the same cities as Office_cities but has the geometries stored in the same projection and coordinate system as the Trails_in_WMA file. We will use this new file for the remainder of the tasks. However, you probably initially observed the other outcome, which is that you can see some of the information from the `Result` object from running the `Project` tool. This is useful if there are problems that are reported within the Messages component of the result:

Output
```
J:\noSpacesBackup\arc\DigitalEarthTest\2013\Lab7Data\Analysis\Office_cities_proj.shp
```
Messages
```
Start Time: Thursday, October 7, 2021 11:40:40 AM
Succeeded at Thursday, October 7, 2021 11:40:40 AM (Elapsed Time: 0.39 seconds)
```

Edit the relevant cell above to make sure the `cityCursor` relates to the new feature class, and rerun it. You should now have distances that are closer to what you would expect. However, the nested for loops make it seem that you should get distances from more than one town. After all, `cityCursor` is looping through five different cities, but only one is analyzed:

```
trailCursor = arcpy.da.SearchCursor("Trails_in_WMA", ["SHAPE@", "TRAIL_NAME"])
cityCursor = arcpy.da.SearchCursor("Office_cities_proj", ["SHAPE@", "DESCRIPT"])
for city in cityCursor:
→ for trail in trailCursor:
→ → print ("The distance from ", city[1], " to ", trail[1], " is ", str(***Answer to Question
10***), " meters")
```

```
The distance from Marlinton town to Durbin to Cass Connector Rail-Trail is 37008.85671598571 meters
The distance from Marlinton town to Blackwater Canyon Rail-Trail is 102933.77112816731 meters
The distance from Marlinton town to Shaver's Fork Valley Trail is 35882.950556587464 meters
The distance from Marlinton town to Long Point Trail is 67815.68038199408 meters
The distance from Marlinton town to Shavers Fork Road Trail is 82691.17533244715 meters
The distance from Marlinton town to Allegheny Trail is 108804.50092157851 meters
```

Question 11: Thinking back to how cursors work, and how this might impact the use of two nested cursors, why didn't you get distances for all five office cities?

In the previous chapter, we addressed this by redoing the `trailCursor = ...` line. As is often the case in programming, that isn't the only possible solution. Cursors come with a `.reset()` method that, as the name suggests, resets the cursor to be pointing once again to the first row in the attribute table.

Question 12: Revise the nested for loops by adding `trailCursor.reset()` at an appropriate line and indentation level.[2] (You can simplify the print statement by writing "print the distance here.")

Question 13: List the closest trail to each city and the distance. (Round distances to the nearest meter.)

Marlinton:

Petersburg:

Reedsville:

Summersville:

Weston:

Task 3: Creating New Geometries with Buffers

Another method for geometries is creating buffers. Just as running some of the raster tools do not create permanent rasters, it is possible to create a buffer and other geometries, in a way that does *not* create a new shapefile or feature class. It instead creates a polygon object for ArcPy's use, but does not turn it into a layer to be drawn onto a map. This is how the buffer method of the `Geometry` object differs from the Buffer tool in the Toolbox or the associated ArcPy command. The advantage is that it will be faster than creating a whole new shapefile or geodatabase feature collection.

To store the areas, we will use a dictionary to match a key (in this case, the trail name) with a value (in this case, the buffer area). The `for` loop after this initialization process will populate the dictionary with the values of the area of a 2000 meter buffer around each trail;

```
trailCursor.reset()
areaDict = {}
for trail in trailCursor:
→ name = trail[1]
→ shape = trail[0]
→ buffer = shape.buffer(2000)
→ area = buffer.area
→ areaDict[name] = area
```

To confirm it worked, enter this line to retrieve the value associated with the Allegheny Trail:

```
areaDict["Allegheny Trail"]
```

Question 14: How big, rounded to the nearest square meter, is the buffer around the Allegheny Trail?

Task 4: Carry Out Geometry Operations with Two Polygons

Our next task is to calculate the ratio of the size of the wildlife management area to the size of the buffer for the trails inside that management area.

---

[2]Think through the indentation, variable, and placement of the reset method carefully, and just to be sure, save the project before you run your revised code—there is a way to get it wrong which creates an infinite loop!

Assume that, for environmental purposes, a trail will require further environmental impact review if its buffer occupies more than 10% of the management area. Therefore, we want to know how much of the management area will be impacted.

First, reset the WMA cursor:

```
wmaCursor.reset()
```

Planning this, we might come to the following recognitions. As in the previous section, we will want to use nested `for` loops with a reset statement. We will need to use an `if` statement to only perform the ratio calculation for trail/WMA pairs where the trail is inside the WMA. You should use one of the geometry methods to identify whether the trail is inside the WMA.[3] In other words, don't bother finding the ratio of the Allegheny Trail buffer area to the Summersville Lake PHAFA area, because the Allegheny Trail is inside the Blackwater WMA. We already have a dictionary to provide the buffer area, so redoing those calculations won't be necessary. For now, we can calculate the ratio as 100 * buffer area/wildlife area, ignoring the possibility that the buffers extend outside the wildlife management area.

These lead to the following code, and use this output to fill in the table for Question 16:

```
wmaCursor.reset()
for wma in wmaCursor:
    wmaShape = wma[0]
    wmaName = wma[1]
    trailCursor.reset()
    for trail in trailCursor:
        trailName = trail[1]
        trailShape = trail[0]
        trailArea = areaDict[trailName]
        if wmaShape.contains(trailShape):
            ratio = 100*trailArea/wmaShape.area
            print(wmaName, " contains ", trailName, " with ", ratio, "% possible overlap")
```

Question 16:

Trail	WMA	Percent Covered

Task 5: Creating a New Geometry from Scratch

Now imagine that you have the planned route for a new trail within a WMA. Using coordinates provided, we will create a geometry for this and add it to the trails layer using an insert cursor.

The first thing to do is create the points, which is simply accomplished through an arcpy function for creating points, aptly named `arcpy.Point()`. We will use this to create seven points for our new trail. These points then have to be formed into

---

[3]There are a few which will work, so don't panic if you take a look at the documentation and come up with a different strategy than I have here.

an array, because that, after all, is how the coordinates of a line are represented. ArcPy provides a function for creating an array, as well, which takes as its input a list of the points you want in the array:

```
point1 = arcpy.Point(599347.2, 4388183.01)
point2 = arcpy.Point(599915.79, 4387982.42)
point3 = arcpy.Point(599715.2, 4387320.48)
point4 = arcpy.Point(600036.14, 4387099.84)
point5 = arcpy.Point(599955.91, 4386377.72)
point6 = arcpy.Point(600457.37, 4386116.96)
point7 = arcpy.Point(600818.43, 4385976.55)

array = arcpy.Array([point1, point2, point3, point4, point5, point6, point7])
```

The other thing we need to create a geometry is to have an idea what its coordinate system is, which is stored in that geometry's spatial reference. After all, we would like ArcGIS to know whether we are dealing with latitude/longitude, UTM, or something else. In this case, as we are adding to an existing feature class, we can find the spatial reference of that feature class, by getting the spatial reference of one of the other geometries for that feature class. Since we have been reusing the geom variable, we need to go back and make sure it is referring to the correct cursor. Once the geometry is set up, we can access its spatial reference just like any other property. In this case, rather than looking at it, we want to store it into a variable for later use. Once this is set up, we can now actually create our new line geometry. There are some other attributes of the trails that would be worth filling in, so we will create variables for these, as well. They are separate variables, rather than used directly in the insert cursor's command, because it is easier to identify what values correspond with which attribute and edit the script properly if needed, by using variables here. Once that all the data are prepared, we can add a feature to our feature class with an InsertCursor:

```
trailCursor = arcpy.da.SearchCursor("Trails_in_WMA", ["SHAPE@", "TRAIL_NAME"])
trail = trailCursor.next()
geom = trail[0]
spref = geom.spatialReference
newLine = arcpy.Polyline(array, spref)
name = "Cheat Canyon Trail"
length = 0
cat = "Proposed Rail-Trails"
stat = "Proposed"
insert = arcpy.da.InsertCursor('Trails_in_WMA', ["SHAPE@", "TRAIL_NAME", "LENGTH", "CATEGORY",
"Status"])
insert.insertRow([newLine, name, length, cat, stat])
del insert
```

Running the code above will probably give a mysterious numerical output. In this case, it is probably a 6, which indicates that we have added a trail with FID #6 to the feature class. To see the new trail, I had to go to the attribute table and select the new row. Then I shifted over to the map, right-clicked on the layer name in the table of contents, went to the Selection section, and chose Zoom to Selection. Once zoomed in on the map, you can use the identify tool on the surrounding WMA to answer the following question:

Question 17: Which WMA contains our new trail?

Task 6: Debugging Exercise

The following code, when it works, will find the intersection of buffers around the trail and the associated WMA so that the ratio can correct for the fact that buffers can extend beyond the boundary of the WMA. If fixed, this could help you with the Challenge Task 8 below. For whatever reason, to create polygons in the Geometry intersect() method, you need a dimension value of 4. There are four errors:

```
wmaCursor = arcpy.da.SearchCursor("WMA_with_trails", ["SHAPE", "NAME"])
ratioList = []
trailCursor.reset()
for trail in trailCursor:
    for wma in wmaCursor:
        if trail[0].contains(wma[0]):
            adjBuffer = trail[0].buffer(2000).intersect(wma[0])
            adjArea = adjBuffer.area()
            ratio = (100 * adjArea/wma[0].area)
            ratioList.append(ratio)
            print(trail[1], "is within", wma[1], "and the ratio is", ratio)
    wmaCursor.reset()
```

Task 7: Unguided Work

Add the state_park_utm and the nps_sept03_utm83 feature classes to the map if they aren't there already.

We want to conduct market research in an area that might be very popular with nature enthusiasts, so the services could be improved. One area was identified as a place that is in both the New River Gorge National River (from the nps feature class) and in Babcock State Park.

Write a script that achieves the following tasks, if copied and pasted into the Python Notebook:

1. Use search cursors to extract these two features.
2. Use the appropriate geometry method to create a polygon of the *intersection* of these two features. Look at the documentation to identify which geometry method to use.[4]
3. Use the `analysis.Select` function to create a feature class called "Babcock" that is just Babcock State Park. (It can be either a shapefile or a geodatabase feature class.)
4. Use the `analysis.Clip` function to clip the Babcock feature class from part 3 to the market area from part 2, and call the output feature class "MarketArea." (Again, either a shapefile or geodatabase feature class is acceptable.)
5. Save to a .py file and add 4 comments—one at each of the appropriate line(s) to describe the code for the four steps above.

Challenge Task 8

This is an extra challenge task: clip the buffers to the WMA areas, and add an evaluation attribute to the trails file.

The ratio calculation I asked you to do ignored the possibility that the buffers might leave the WMA. Mapping it out, most of them do.

Revisit the code in Tasks 2 and 3 to account for this, and only use the area of the buffer that is also within the WMA.

Also add an attribute to the trails feature class called "Evaluation" with a Boolean type, containing `True` if it can be built and `False` if it cannot, using the criterion set out at the start of Part III.

Hints

Look through the geometry section of the ArcPy help documentation to find potentially useful methods.

This may require a more thorough revision that you initially anticipated. That's why it is an extra challenge task!

---

[4]If you access a tool from ArcToolbox instead of the geometry method for accomplishing this, you will lose points.

Introductory Comments: Using UML and Class Diagrams to Manage Complex Objects

We saw in the previous chapters how to investigate, how to manipulate, and, to a lesser extent, how to analyze the data structures of our GIS data. These data structures are not just intertwined with analysis and geoprocessing but are also heavily involved with cartography. There is a lot of computation behind the scenes that ArcGIS carries out within the framework of computer graphics, and these also relate to the data structures. It has to take spatial coordinates, like a latitude-longitude pair, and figure out where on the screen that pair is located. Then there is simplification and generalization of vector data, in which the lines and polygon borders are potentially simplified to make drawing them faster. This is especially critical for small-scale (zoomed-out) maps. If your line for the Monongahela River is very detailed, tracking all the small twists and bends, but the map is zoomed out to the entire state of West Virginia, it is probably the case that there are entire bends contained within a single pixel on the screen. If so, there is no need to attempt to draw all the details, when it will make no difference in how the details appear on the screen. The Douglas-Peucker line generalization algorithm is a way of carrying out this simplification.

There are similar approaches to simplifying raster data to make it faster to load and render. This is through pyramids, which you have probably indirectly encountered, because of how often ArcGIS asks whether you want to create pyramids. These are representations of the data at different resolutions, merging pixels, because, like the bends of the river, there is no need to try to directly represent the raster pixels individually if many of them will fit within a single screen pixel.

All algorithms for these computer graphics depend upon the GIS data structures and, in a way, create new GIS data structures, even if they are temporary ones for displaying the data on the screen. We will not go into the specific details of these computer graphics concerns here, because ArcGIS handles them for us. We will, instead, look at another approach to data structures in computing. While it isn't the kind of data that you directly put onto a map, there is a lot of information contained in the representation of how the map is constructed from the GIS data. This extra information, like how many layers are in the map, what color a certain feature class is, and the class breaks for a choropleth map, also all need to be represented within the computer. This cartographic information takes the form of another series of computational data structures. This chapter and the next one look at these computational data structures within the arcpy mapping module (called `arcpy.mp` in ArcGIS Pro—if you see `arcpy.mapping`, you are looking at a script for ArcGIS Desktop, not Pro).

These computational data structures for cartography again use classes and objects. While we have seen and worked with classes and objects before, the mapping module uses a wider variety of classes than we have encountered thus far. This is because there are many steps that ArcGIS takes when going from, say, a map containing multiple layers all the way to drawing the overlook within the state park as a bright red dot with a diameter of four screen pixels within a light green polygon having a darker green border of a thickness of two screen pixels. Each of these steps provides access to a different set of information and decisions about how to draw the image and what to draw on the screen. Just as we took several steps in Chap. 5 to go from a file path pointing to a CSV file to an object that can read the CSV file, and had a different variable of a different type for each step along the way, each of these cartographic steps also has its own object and class. To keep sense of all of these classes, we will use a fuller version of UML class diagrams than those introduced in Chap. 7. These diagrams show not only the general content of each class but also how classes are connected to each other. The arrows show which class is used

---

Supplementary Information The online version contains supplementary material available at [https://doi.org/10.1007/978-3-031-08498-0_12].

to access other classes. Methods and properties that are used to connect the classes shown are included, although other properties and methods are not included to conserve space and clarity. The numbers tell you how many instances of the subsequent class can be accessed. For example, the `listMaps()` method of the `ArcGISProject` class gives you a list of maps, which could contain any number of maps, including zero, so this is labeled with 0…n. A fuller description of UML diagrams like this is provided in the Appendix (Fig. 12.1).

The first step is the ArcGIS Project, represented in a class called `ArcGISProject`. Through the methods we have seen before, we can access the list of maps from the project. We can also access a list of layouts, which is the subject of the next chapter. Each map, then, has a list of layers, which may or may not be empty. This much, we have seen before. I could extend from the Layer class to some of the data-oriented properties and methods, such as `dataSource`, but as this is a chapter on cartographic symbology, we continue with the Symbology object, accessed through the `Layer.symbology` property.

The Symbology object is equivalent to the Symbology panel in ArcGIS Pro, accessing the same information. Just as there are different versions of the symbology panel for rasters and vectors, we can use an object of the Symbology class to access either a raster colorizer through the `Symbology.colorizer` property or a vector renderer through the `Symbology.renderer` property. The tasks in the remainder of this chapter will only use the renderer, although colorizers are discussed below to introduce you to them.

Furthermore, we can see that there are two types of raster colorizers which are supported by arcpy. One, `RasterUniqueValueColorizer`, is for (as the name tells you) using a unique value for each distinct value, or a set of values, while the other, `RasterClassifyColorizer`, is for stretched values. Both of these use a color ramp, accessed in both cases through a `colorizer.colorRamp` property. You will also see next to both the colorizer subclasses that there are properties with more basic Python types simply listed there. For example, `RasterUniqueValueColorizer` has a property called `field` which is of type String. This is the field used to determine the colors which are shown. Similarly, `RasterClassifyColorizer` has a `classificationField` property of type String. In addition, it has a `classificationMethod` property, with options like "NaturalBreaks" and "Quantile" that tell ArcGIS how to carry out the classification.

Another distinction between the colorizers is that, while both use a color ramp to get the colors, the `RasterClassifyColorizer` has a list of class breaks, each of which is of the class `RasterClassBreak`. These, in turn, store a representation of the color as a dictionary. The details of how a dictionary becomes a color will be seen later, in the discussion of vector symbologies. There is an upper bound to the class, which is a double type. There is no lower bound, however, which suggests that the class breaks are sorted, and the upper bound of the previous class can be considered the lower bound of this one. The label is represented as a String, and this is what the class is called in the legend and in the Table of Contents. Meanwhile, the `RasterUniqueValueColorizer` doesn't have a set of class breaks, because this is not relevant to representing data by their unique values. Instead, there is a list of `ItemGroup` objects, accessed through the `RasterUniqueValueColorizer.groups` property, and each of these item groups has itself a list of items, and each of these items associates a color with a list of values.

This illustrates the number of steps needed to change the color of forest pixels in a land cover map. You have to get the project and access the correct map and then the correct layer within that project. Then you get that layer's symbology and, as a raster layer, its colorizer. Since a land cover map would be using a unique value colorizer, access that colorizer's item group, followed by the appropriate raster item for which "forest" is in the values List. Then you can change the color of that item.

Unfortunately, vector symbology isn't any more streamlined (Fig. 12.2). Vector symbology at least initially appears even more complicated than the raster symbology, although there are actually some similarities. Just as there are two types of colorizers, there are four varieties of renderers. The `UniqueValueRenderer` is, unsurprisingly, similar to the `RasterUniqueValueColorizer`. Both have a `groups` property that accesses a list of objects of class `ItemGroup`. The difference is that while the `items` property of the `RasterUniqueValueColorizer` gives you a list of `RasterItem` objects, the `items` property here gives a list of objects of the type `Item`. Both the `Item` and `RasterItem` classes have a list of values, although whereas the `RasterItem` associates all the values in this list with a color, the `Item` associates them with a `Symbol`. The symbol is, in a sense, the heart of vector symbology. Most renderers access a symbol somehow or other. We will look at symbols in more detail shortly. Also, just as the `RasterUniqueValueColorizer` accesses a `ColorRamp` object, so does the `UniqueValueRenderer`.

Where there is a `RasterClassifyColorizer` to represent classified data within a raster, there are two main ways of representing classified data through this symbology: by altering the colors and by altering the symbol sizes.[1] The `GraduatedColorsRenderer` is the approach of altering the colors, while the `GraduatedSymbolsRenderer` is the approach of altering the symbol sizes. The properties match this distinction, yet there are differences brought on by those

---

[1] Yes, I am aware of dot density maps and other approaches, but they aren't covered here.

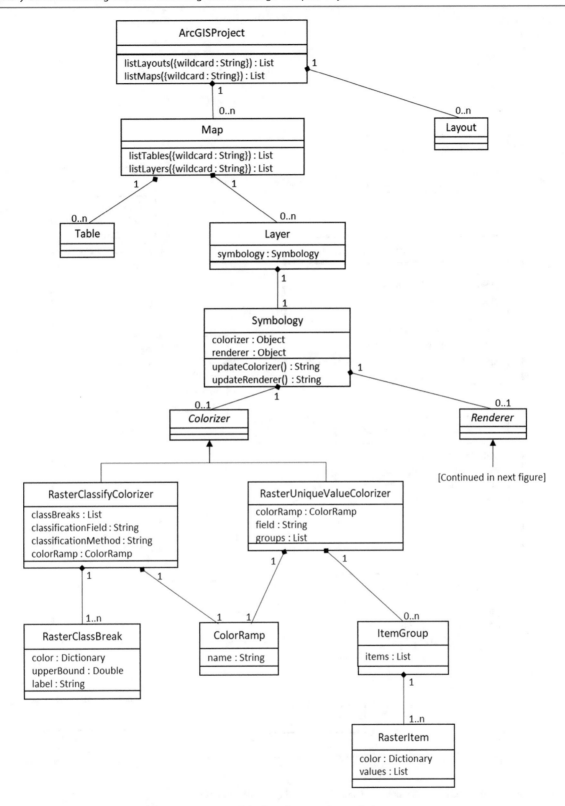

Fig. 12.1 Class diagram for a portion of the `arcpy.mp` module, focusing on raster symbology

approaches. Both have a list of `ClassBreaks`, just as the `RasterClassifyColorizer` does. The `ClassBreak` and the `RasterClassBreak` are very similar, as both have a label for how it will appear in the legend, both have an upper bound denoting the top range of the class break, and both assume the lowest value starts the lowest class and the upper bound of the preceding class always sets the lowest bound for this one. The difference is that the `ClassBreak` represents each

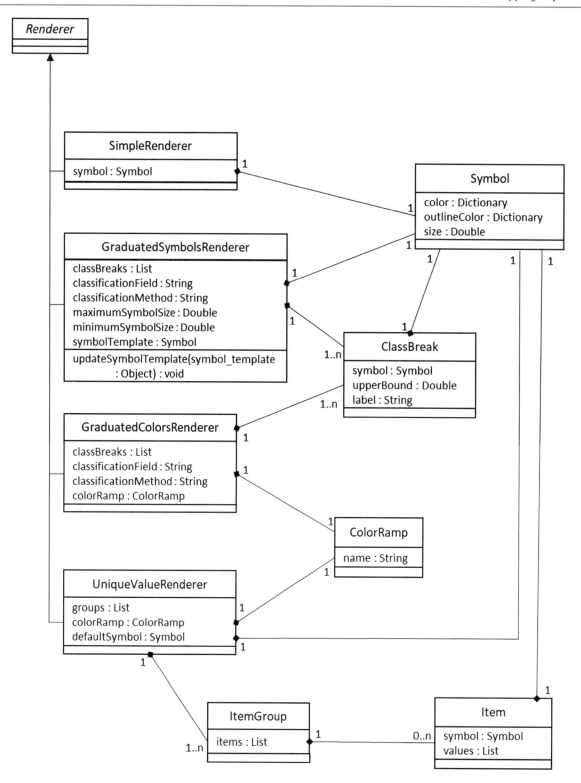

Fig. 12.2 Class diagram for a portion of the `arcpy.mp` module, focusing on vector symbology

class by its symbol, while the `RasterClassBreak` represents it by the color, since symbols do not apply to raster data. However, because of its nature of classifying by color, the `GraduatedColorsRenderer` has a `colorRamp` property, while the `GraduatedSymbolsRenderer` has properties for minimum and maximum size and a `symbolTemplate` property which represents all aspects of the symbols except the size.

The renderer type which has no analogue within the raster colorizers is the `SimpleRenderer`. After all, while a simple dot map with all cities represented with the same symbol is a potentially useful map layer, a raster with all pixels simply colorized to the same color is not very useful. Therefore, there's a `SimpleRenderer` that just has a `symbol` property.

This `symbol` property provides access, as do other properties of other renderers, to an object of the `Symbol` class. This represents how the feature is actually drawn on the map, so if you want to, for example, change the size of a point, or the color of a polygon, you have to do so through the `Symbol` that is being used to draw that object. There is a lengthy series of properties belonging to the Symbol class, only a couple of which are shown in the diagram. The colors are, as with the raster colorizers, represented as dictionaries, which might seem an unusual choice. There are several ways of representing colors in computer graphics, so just assuming that your colors are going to follow the most common format of RGB (red, green, blue), is not a safe assumption. Therefore, `arcpy` gives you the opportunity to represent colors in one of several different ways through a dictionary with exactly one entry. This entry uses the color format (RGB, CMYK, HSV being common ones) as the dictionary item's key, as a String. The value associated with that key is a list of numbers, such as the red, green, and blue values ranging from 0 to 255. There is, furthermore, an additional value, called alpha in the documentation. This impacts the opacity or transparency, with 100 as completely opaque or not at all transparent and 0 as completely transparent.

The following tasks introduce some of these symbology representations through vector renderers. It is not a complete enumeration of everything within `arcpy.mp` by any stretch but may serve as a useful introduction to scripting with symbologies.

Task 1: Basic Mapping Tasks—Adding Layers

Throughout this lab, we will be working with a geodatabase of some reference layers: counties (USCensus2010_arc), states, highways, rivers, and cities. The overarching goal will be to produce a map of county by income with the other layers there for reference. Copy everything over to your preferred workspace and open the Python Window, as we will use that for the guided tasks in this chapter.

I have given you a basic starting project already, so open mapping_start.aprx, and fix the inevitable broken data references. You will notice that one of the layers listed above is missing: the rivers. We will start by adding this layer using scripting. Open the Python Window and enter these two lines:

```
aprx = arcpy.mp.ArcGISProject("CURRENT")
thisMap = aprx.listMaps()[0]
```

Question 1: Why might it be better to set the map to "current" rather than designate a file location?

Now we will add the rivers layer. To do this, we want two layer objects: one for the layer we are inserting into the document and one that we can use to tell ArcMap where in the Table of Contents to put it. The `listLayers` function works similar to the `listMaps` function, but here we are specifying the name we want:

```
roadLayer = thisMap.listLayers("roads")[0]
```

Question 2: Why would specifying the parameter for "roads" be better than using the index of the roads layer (e.g., `thisMap.listLayers()[1]`)?

As with the `listMaps` function, it returns a list, even when there is only one item in the list, so the `[0]` is necessary. We have to point ArcMap to the right geodatabase layer to give it the rivers. (Edit the file path as necessary.) Once complete, it adds the layer to the map and returns a Result object:

```
>>> riverLayer = arcpy.management.MakeFeatureLayer("C:/noSpaces/arc/WVAGP/mapping/mapping_start.
gdb/rivers", "river_lyr")

<Result 'river_lyr'>
```

Task 2: Changing Layer Properties

It may not be very useful to us to have the states covering up the counties like this, so let's make the states layer invisible. To retrieve the states layer and put it into a variable for continued access, enter the following line:

```
statesLayer = thisMap.listLayers ("states")[0]
```

Take a look at the properties of the Layer class by going to the arcpy for ArcGIS Pro help website and search for Layer class. (It's also at http://pro.arcgis.com/en/pro-app/arcpy/mapping/layer-class.htm.) Keep this window open; it will be useful later! These are the attributes of the layer that you can change through arcPy. Some of them are "Read Only," like `isRaster-Layer`. This means that arcPy will be able to tell you whether or not it is a raster layer, but you won't be able to change it. This makes sense, as it would create major problems and probably cause ArcGIS to crash if you could change a layer from a vector layer to a raster layer without concurrently switching the dataset that it refers to from a vector to a raster. Others, like `transparency`, are labeled "Read and Write," meaning you can use the code to change a layer's transparency. Do any of the properties look particularly useful for what we want to do? Try this line:

```
statesLayer.visible = False
```

Hit enter. You should have a map like this, with the box next to the name of the states layer unchecked.

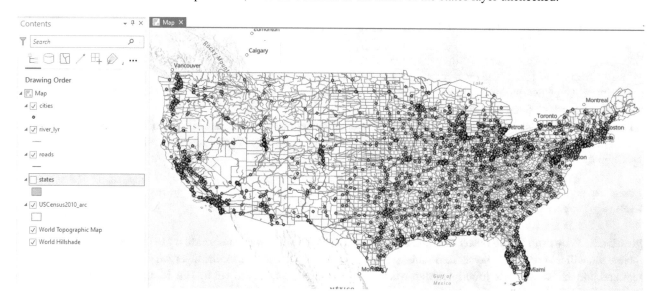

So we've turned off the states to see the counties, and this can be worthwhile. But maybe we'd prefer to make the states hollow with a thicker outline and give them labels, for reference purposes. We will first tackle the labels and address the symbology in the next task.

Question 3: What code can you use to make the states layer visible?

Now let's add labels to the states. There should be a property of the Layer class that you can find which will allow you to display the labels.

Question 4: What is it? What code do you need to include here?

Just as there are different types of datasets that we examined with the `Describe` object, there are different types of layers (raster, feature, etc.), and again analogous to the properties of the `Describe` object, some of the layer properties are only available to certain layer types. We can check whether or not a property is supported using the `supports()` method.

For example, this line checks whether `statesLayer` supports the *contrast* property. Then create variables to access each layer, not just the states and road layer variables that you have at the moment.

```
statesLayer.supports("CONTRAST")
```

Question 5: Create variables to access each layer, and use them to identify which layers (if any) in our geodatabase support the *brightness* property? Enter the relevant code here.

Task 3: Changing Symbology

As stated above, we would like to make the states hollow. This can be done through the symbology property. As described in the introductory remarks, there is a chain of properties we will need to go through to edit a symbology. The first of these is the symbology property of the layer. That, in turn, has a renderer, which records how the layer as a whole is drawn. There are several types of renderers, for different ways of displaying the layer. Find out what the renderer type of the states layer is by typing the next lines and hitting Enter:

```
statesSymbology = statesLayer.symbology
statesSymbology.renderer.type
```

You should get the answer to Question 6 below:

Question 6: (5 points) What is the renderer type for the states layer?

This type of renderer means each feature is displayed using the exact same symbol, which we can access through the first of the following lines. This makes sense, as looking at the map, all the states will be the same color with the same outline thickness. (The exact color may vary depending upon how ArcGIS Pro seems to feel that day.) We wanted it hollow, which is through the second of the following lines:

```
statesSymbol = statesSymbology.renderer.symbol
statesSymbol.color = {'RGB' : [0, 0, 0, 0]}
```

Question 7: Using the help documentation for the symbol class, why is the *last* number in the list specifying the color a 0?

You should not see any changes. What we have done so far is edit the variable holding the symbology for the layer. Think back to the variable called bob in Chap. 2, which was given the item in a list, but changing the variable bob didn't change the list. A similar dynamic is happening here, in which the `statesSymbology` variable was changed, but that didn't change the symbology property of the layer.[2] We now need to go back to save those edits into the layer. This is done through setting the symbology property of the layer. (The symbology property is, after all, both read and write enabled.)

```
statesLayer.symbology = statesSymbology
```

Now it is hollow, but to make the state borders distinct from the county borders, we want to give it a thicker outline. This is through the `size` property:

```
statesSymbol.size = 3
statesLayer.symbology = statesSymbology
```

---

[2]I say "similar dynamic" instead of "same dynamic" because, for some reason, editing the variable called statesSymbol *did* automatically have the changes get reflected within the statesSymbology variable, even though changes to statesSymbology have not yet been reflected by the statesLayer variable. To use computer science jargon, it is as if ArcPy used call by reference to access statesSymbol but call by value for accessing statesSymbology.

Now let's move on to the counties. We may first want to symbolize counties by MedianAge, with a graduated color scheme. You should already have a county layer variable created as part of your process for Question 5. If the variable name is different from `countiesLayer`, substitute the appropriate name. The first step is accessing its symbology. Next, we want to change the renderer to support graduated colors. This is through the `updateRenderer()` method of the symbology object. As before, you'll need to then take the altered symbology variable and put it back into the counties layer object's symbology property to make any changes to the map. When done, you probably have a choropleth map of land area (ALAND10) using a yellow to red color scheme:

```
countiesSymbology = countiesLayer.symbology
countiesSymbology.updateRenderer('GraduatedColorsRenderer')
countiesLayer.symbology = countiesSymbology
```

Now that we have changed the renderer type to a choropleth map, this particular choropleth map might not be what we want. Therefore, we can use scripting to make some changes to this. First is the attribute we are using. Then we can change the color scheme to shades of blue, if we want. This is through the list of available color ramps through the `aprx` object. Lastly, we again want to make sure the changes are reflected in the map:

```
countiesSymbology.renderer.classificationField = "MedianAge"
countiesSymbology.renderer.colorRamp = aprx.listColorRamps("Blues (5 Classes)")[0]
countiesLayer.symbology = countiesSymbology
```

Task 4: Debugging Exercise

Let's say that we want to use a different set of class breaks, as well as a different set of colors. When the bugs are fixed, the following code will change the counties layer to use a quantile break point. They will also use shades of gray with the following colors, all by the RGB scale: (30, 30, 30), (60, 60, 60), (100, 100, 100), (150, 150, 150), and (200, 200, 200). There are four bugs:

```
colorsList = [[30, 30, 30, 100], [60, 60, 60, 100], [100, 100, 100, 100], [150, 150, 150, 100],
[200, 200, 200, 100]]
countiesSymbology = countiesLayer.symbology
countiesSymbology.renderer.classificationMethod = 'Quartile'
countiesSymbology.renderer.breakCount=5
breaks = countiesSymbology.renderer.classBreaks
index = 0
for break in breaks:
 break.symbol.color = {'RBG': colorsList[index]}
countiesLayer.symbology = countiesSymbology
```

Task 5: Unguided Work

You will now use the help documentation, the preceding tasks, and potentially other available resources to change the symbology of the map to meet the following specifications. Use arcpy, instead of the GUI, and stay in the Python Window so that changes are immediately visible.
Cities layer:
(5 points) Change to a size 2 pt circle. You may leave the color the same.
River layer:
(5 points) Change it so the rivers are blue (any shade of blue is acceptable).
Roads:
(5 points) Change it to use the built-in "Highway" symbol
States:

Leave the states layer as is.

Counties:

Change it to be a choropleth map using:

(5 points) the attribute of per capita income (PCIncome)

(5 points) with 7 classes

(5 points) an equal interval classification system

(20 points each, 30 total) using the following labels and colors (in RGB format)

Label	Color
Very low income	(0, 20, 0, 100)
Low income	(0, 60, 0, 100)
Moderately low income	(0, 100, 0, 100)
Middle income	(0, 140, 0, 100)
Moderately high income	(0, 180, 0, 100)
High income	(0, 220, 0, 100)
Very high income	(0, 255, 0, 100)

Mapping: Layouts

Introductory Comments: Data Structures for Maps Versus Layouts

This chapter follows on from the previous chapter, but where Chap. 12 looked at the map figure and how to make changes to the appearance of layers in the map, this chapter looks at the layout of the map on the page. In terms that would be familiar to ArcGIS Pro users, fully aware of the distinction between the map view and the layout view, Chap. 12 examines what is possible in arcpy with respect to the map view, while Chap. 13 here examines what is possible in arcpy with respect to the layout view.

As with the classes in Chap. 12, visualizing the structure of the relationships among the classes can help, and this is presented in Fig. 13.1. As before, we have the Map and the Layout classes accessed through the ArcGISProject. Where Chap. 12 looked at classes attached to the Map, this looks at classes attached to the Layout. They are all within the framework of map elements and accessed through the `.listElements()` method. There are several types of map elements that can be contained within this list, and the tasks below use four of them: a map frame element, a legend element, a map surround element which refers to things like scale bars and north arrows, and a text element.

While there is no "LayoutElement" superclass that could hold these properties, there are a few properties that are common to all of the individual element classes (even including those not presented here). They all refer to the size and position of the element as it relates to the page: the height, width, and X and Y coordinates. There are other properties and methods specific to some of the types of Layout element that we are using here. The map frame is, by its name, the element that holds the map figure. This is why it has a reference to the map inside the frame through the `.map` property and a reference to a camera object, which is telling the map frame where on that map to center itself and what scale to use for zooming in or out.

The map frame is referenced by the next type of element, which is the legend element. After all, the legend provides a visual description of the contents of a map, so it needs to be able to refer to a map frame to access the contents it is describing. It also has a list of LegendItems in which those descriptions are held. The order of the contents of a legend item is in the arrangement property.

The remaining two layout elements that we will be working with in this chapter are the most straightforward ones. Text elements contain text and, as such, refer to things like the map title or the credits. The nature of a map surround element is not as evident by its name, but it refers to the other ancillary elements of a map, like a north arrow and a scale bar.

The following tasks introduce you to and have you work with some of these elements for the purpose of improving the cartographic presentation of the data. As a preliminary note, it is worth acknowledging the limitations of arcpy in this regard. It is possible to move elements on the map and, in some limited cases, edit them. However, there is no way strictly within arcpy to create a map element, whether it is the map frame, the legend, the north arrow, or anything else. Therefore, we will be using a mix of the GUI and the Python Window for this exercise. It is still useful to employ arcpy, even within this limited setting, because it can allow you to exercise very fine control over the layout and be able to record the exact placement of each element within the layout so that it can be recreated with the same or different data.

J. Conley, *A Geographer's Guide to Computing Fundamentals*, Springer Textbooks in Earth Sciences,
Geography and Environment, https://doi.org/10.1007/978-3-031-08498-0_13

Fig. 13.1 Class diagram of
selected Layout-related
classes

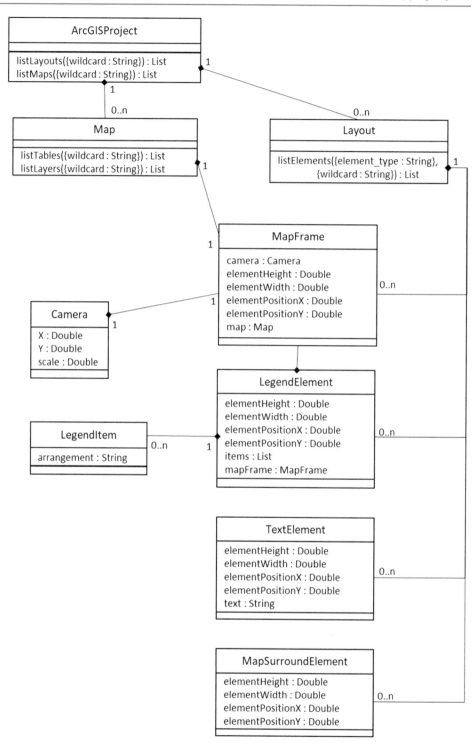

Task 1: Working with the Map Frame

Open up the project from Chap. 12 and have it as you did at the end of the unguided portion.[1] Use the GUI to add a layout to the project. This is through the Insert tab ➔ New Layout. Use the letter size at the top left of the options that are presented.

Once you have the layout open, we can begin to access it with arcpy commands. Just as we always worked with lists for the map, we also work with lists here, even when there is one layout in the project.

Enter the following code into the Python Window:

```
aprx = arcpy.mp.ArcGISProject("CURRENT")
layout = aprx.listLayouts()[0]
```

Most of the time, we will then use the layout to access various elements within it. This is through the following line:

```
layout.listElements()
```

Question 1: How many elements are currently in the map?

Even with this layout as generic as it is, we can see some useful information from it, as it stores the page information. Especially important is to recognize the page units, because we are introducing what is, in a sense, a second coordinate system: the coordinates of the page. These are what will be used to locate and place items on the page. You can see in the layout view that there are rulers along the top and left sides of the layout window. These represent the page coordinates. Using either the documentation or the Python Window strategy of waiting after the dot to see the property and method options, answer the following questions:

Question 2a: What is the height, width, and units for the page?
Question 2b: What code did you enter to find this out?

As our layout isn't very useful at the moment, the first thing we would want to do with it is to put a map frame in it. We will have to create a map frame in GUI. To do so, have the layout view active, and then go to the *Insert* tab and choose *Map Frame*. Then drag a rectangle in the layout to place the map frame. You do not have to be precise in your drawing of the rectangle because we will be using arcpy to make its position and size exactly where and what we want them to be. Once you do so, you might have a layout like this (depending, of course, upon your zoom level for the project and the exact rectangle you drew).

---

[1] Strictly speaking, it isn't mandatory to have it exactly as this was instructed at the end of Chap. 12. The layout code will work no matter what stage of Chap. 12 it is on, although the legend size in particular might seem off if you have a different number of layers visible or a different number of classes in the choropleth map.

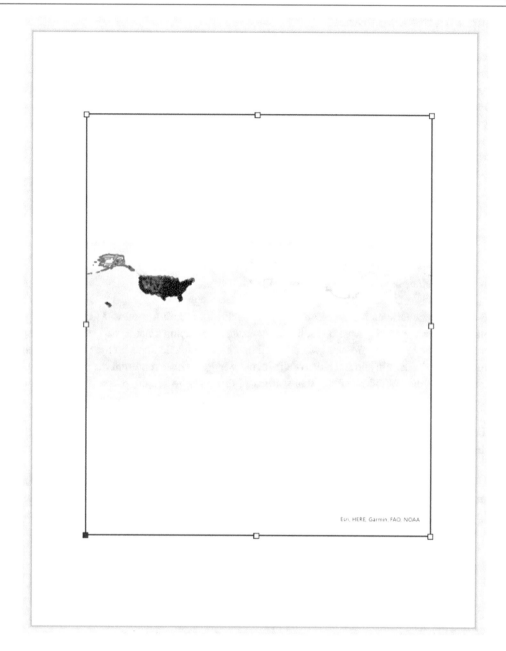

While this is a decent start, there is a lot of work to do. To accomplish that work, we need access to the map frame we just added. Enter `layout.listElements()`, and press enter to confirm that there is now an element within the layout. This element is the map frame, so we can access it through the following line:

```
mapFrame = layout.listElements()[0]
```

It is sensible to then want to ensure the zoom level of the map is where we want it, as well as ensuring it is centered properly. The map above fails both of these criteria. To remedy this, we need to access an intermediary object, the map frame's camera. It is the camera that records the scale, the position, and the extent of the map within the map frame. Enter the following line to store the camera in a variable:

```
camera = mapFrame.camera
```

We might first want to set the camera's X and Y coordinates, which is what is found in the center of the map. Use the following lines to accomplish this:

```
camera.X = -96
camera.Y = 40
```

You will notice that these coordinates are latitude and longitude, rather than page coordinates, because they refer to placement of the map itself within the frame, rather than the placement of the map frame within the layout page.

It is also probably zoomed to the wrong level. This is through the scale property of the camera. Enter the following line to change it to a scale of 1:40,000,000:

```
camera.scale = 40000000
```

Now that we have the map set up how we want, the next part of the task is putting it in the right part of the layout. This is where the page coordinates come into play. Page coordinates are structured such that the origin is the bottom left of the page, and we measure up from there and to the right from there. We want the lower left corner of the map frame to be 1 inch from the left side of the page and 4 inches up from the bottom. This is done through the element properties of the mapFrame variable. Likewise, we can use element properties to set the precise height and width of the map frame to be 5 inches and 6.5 inches, respectively. Enter the following code to accomplish this:

```
mapFrame.elementHeight = 5
mapFrame.elementPositionX = 1
mapFrame.elementPositionY = 4
mapFrame.elementWidth = 6.5
```

Lastly, recognizing what is portrayed in the map frame, you will want to change the name of the map frame to "Lower48." This is through the following line:

```
mapFrame.name = "Lower48"
```

Task 2: Working with Map Surround Elements

The next portion of the layout we will work with is the map surround elements. There are two that we will use here: the north arrow and the scale bar. In both cases, we will use the generic layout element properties to be specific about the position where we want them, although this is only possible after we add then through the GUI.

First, we will insert the north arrow. While the layout view is active, go to the Insert tab, and select north arrow. Click on the map to place it. As with the map frame, you do not need to be exact, as we will be adjusting the position on the page using arcpy.

Once it is added, we can access it using the input parameter option for the layout.listElements() method. There are several different types of layout elements available in arcpy, and we often know which element we want to work with. Thus, we can specify we want a Map surround element through the following lines. The north arrow has a fixed aspect ratio, so we really only need to set the width, which will be a quarter inch. We then want the north arrow to be below the map frame, on the right side:

```
layout.listElements()
arrow = layout.listElements('MAPSURROUND_ELEMENT')[0]
arrow.elementWidth = 0.25
arrow.elementPositionX = 7
arrow.elementPositionY = 3.25
```

Second, we will insert the scale bar. Again, this is through the Insert tab, but select scale bar, and likewise click on the map. Access the scale bar through the following code:

```
scale = layout.listElements('MAPSURROUND_ELEMENT')[0]
```

You'll notice that except for the variable name, this is the same code as was used to access the north arrow, down to the list index. All items in the list are indexed in the order shown in the "Drawing Order" part of the Contents panel at the left side of the screen, with the top item being index #0 to the end of the list being at the bottom. When we inserted the scale bar, it was added to the top of the drawing order, so it is the new index #0 in the list.

Question 3: Thinking back to how variable assignment works in Python, does the `arrow` variable refer to the north arrow or the scale bar? Why is this the case?

Like the north arrow, we only need to set the width of the scale bar and its X and Y position, which we will want below the map frame on the left:

```
scale.elementWidth = 3.75
scale.elementPositionX = 1
scale.elementPositionY = 3
```

Task 3: Working with Text Elements

For our text, we will want two pieces of text: a title and some simple credits. As with the map surround elements, we need to first create them using the GUI. To insert the title, go to the Insert tab, and in the Graphics and Text section, choose the button with the plain A, labeled "Straight text" if you hover over it. Click on the map to place it. As with the map surround elements, we can use the input parameter to the `listElements` method to only retrieve the text elements. Currently, of course, we only have one. The title should be at the top of the layout, and we want it to be fairly big, so enter the following lines to position it within the layout:

```
title = layout.listElements('TEXT_ELEMENT')[0]
title.elementHeight = 0.5
title.elementPositionY = 9.75
title.elementPositionX = 1.1
```

In addition to the placement, we want to change the text from the generic "Text" that is created by ArcGIS Pro when we put it on the layout. This is through the `.text` property of the `title` variable:[2]

```
title.text = "Reference and Income Map of U.S."
```

Because this currently has the generic title of "Text" in the Drawing Order, we will want to make sure it is obvious which text is the title and which is the credits by changing its name, too:

```
title.name = "Title"
```

The other item to add is the credits. You can insert another text box for this like the title or use the Dynamic Text option by going to the Insert tab and then the Dynamic Text drop-down menu and choose Credits. Then click on the map. Either way, you will get a place for text on the map. We want this text to be at the lower right end of the layout and set its name to "Credits." For the title, we specified the size by setting the height of the text box. In this case, we might be more interested in the size of the text font. We can use the `.textSize` property to set the font size. In this case, a font size of 12 points is good. We want the text on two lines, which means we will make use of the newLine character ('\n'). This is the same thing that may have tripped you up if you didn't do the slashes correctly in file paths, but this time, we want it. Insert the following text, making sure to replace my name with yours:

---

[2]For reasons I don't understand, I had to enter this line twice for it to show up in the layout view. Don't panic if you have to do the same.

```
credits = layout.listElements('TEXT_ELEMENT')[0]
credits.name = "Credits"
credits.elementPositionY = 1
credits.elementPositionX = 6.5
credits.textSize = 12
credits.text = "Guide to Programming \n Jamison Conley"
```

Task 4: Working with Legend Elements

The last element we will want for our layout is the legend. As with the other layout elements, we need to create the legend and edit it within the GUI before working with it in arcpy.

Go to the Insert tab and select Legend, and then click on the map. It probably appeared as a single column that was far too long to be useful within our layout. It will be better with two columns. Now the legend should be a reasonable dimension for what we want, and we can place it in the lower left portion of the layout:

```
legend = layout.listElements('LEGEND_ELEMENT')[0]
legend.columnCount = 2

legend.elementPositionX = 1
legend.elementPositionY = 2.5
```

The other thing we can do is to change the display name of the county layer to be more informative for the legend. To accomplish this, we will need to go back and access the map itself, as the layer is accessed by the map, not the layout. The asterisks before and after Census in the code below indicate what are called wild card characters. They mean we will access any layer with "Census" in its name, whether or not it is the entire name:

```
map = mapFrame.map
counties = map.listLayers("*Census*")[0]
counties.name = "2010 Income"
```

Once all of this is completed, we will lastly want to export the layout. Layouts have many export methods, each one based upon a particular output file extension. We might want to have a PDF for broad shareability, so enter this last line, changing the file path to whatever you have on your machine:

```
layout.exportToPDF("J:/…filePath…/USIncome_layout1.pdf")
```

Question 4: This PDF will be assessed for accuracy of the position of all parts of this layout.

Task 5: Debugging Exercise

To improve the layout further, you might want to have the choropleth legend portion split over two columns. When fixed, the following code should put the entirety of the choropleth legend item in the second column. You may want to examine the ArcPy documentation for relevant classes. There are three bugs, although you might need to make four changes, depending upon the behavior of ArcGIS Pro during the testing and debugging process:

```
legend.fittingStrategy = 'ManualStrategy'
items = legend.items()
choroplethItem = items[-1]
choroplethItem.column = 1
```

Task 6: Unguided Work

You might notice the absence of Alaska and Hawaii in our layout. We can remedy this by adding two inset maps, which are simply additional map frames. Please provide code for all of the aspects of this which can be completed in arcpy.

Move the center of the map to the south to make room for insets by centering it on 35 Degrees North, instead of 40.
Insert two new map frames (using GUI).
Name one Alaska and the other Hawaii.
Alaska gets the following specifications:

- Map X = −150 degrees.
- Map Y = 60 degrees.
- Width = 2".
- Height = 2".
- Scale = 1:80000000.
- Element Position X = 1″.
- Element Position Y = 4″.
- Add text (using GUI) and set the text to "Alaska".
- Text position X = 1.75".
- Text position Y = 5.8".
- Text size 11 font.

Hawaii gets the following specifications:

- Map X = −158 Degrees.
- Map Y = 20 Degrees.
- Width = 1.5".
- Height = 1.25".
- Scale = 1:20000000.
- Element Position X = 3″.
- Element Position Y = 4″.
- Add text (using GUI) and set the text to "Hawaii".
- Text position X = 3.75".
- Text position Y = 5.05".
- Text size 11 font.

Export as USIncome_layout2.pdf.

Part III

Creating Algorithms to Enhance ArcGIS Pro

Once you have the data structures in your GIS, we can combine this with the basics of programming, especially the control statements and the use of functions, to develop new algorithms to extend and enhance the abilities of ArcGIS Pro. After all, most people engage in the programming process for ArcGIS Pro with one of two goals in mind: having scripts to automate one or more processes in ArcGIS Pro so that it is less cumbersome than constantly using the GUIs provided within the toolbox, or writing programs to create a new tool to add functionality that isn't available in ArcGIS Pro.

For either of these, an understanding of algorithms will greatly help the development of the scripts and programs that are needed to accomplish these objectives. While a full exploration of GIS algorithms would be more extensive than can fit into this book, and is already the subject of existing books, this section provides a foundation in the development of algorithms, and places it within the context of creating new tools and new classes for use within ArcGIS Pro.

Chapter 14 introduces the creation of new tools that allow you to run your scripts within ArcGIS Pro directly, and Chap. 15 adds the more sophisticated handling of errors to this to make the tools more user friendly. Chapter 16 extends this to show how to create your own custom classes that can enhance the data structures you use in your programs and scripts. Lastly, Chap. 17 provides an example of the development of algorithms to solve a problem and asks you to develop your own algorithms to solve problems within an applied context.

I will note here that there are two types of custom tools that can be made within ArcGIS Pro. This section focuses on the type called script tools. These are simpler, and therefore a gentler introduction to how scripts can be used to create new tools. The other type, which is introduced in Chaps. 20 and 21 of Part IV, is called a Python toolbox. This approach has more flexibility, and allows for projects that incorporate multiple Python files into one tool. For those of you who get interested in more in-depth development of GIS software, including contributing to the open source QGIS, this more advanced approach may become more useful for you.

Introductory Comments: Concepts of User Interfaces

We have seen thus far the ability to enter Python lines and pieces statement by statement into the Python Window or block by block in the Python Notebook in ArcGIS Pro. Additionally, we have seen how to run standalone scripts within IDLE. However, what if you want to run an entire script within ArcGIS Pro? This is accomplished through custom tools.

Custom tools can serve two purposes. The first is running a script that provides a tool or analytical technique that is not yet implemented within ArcGIS Pro. For example, some forms of spatial regression, like spatial lag regression and spatial error regression, are not implemented within ArcGIS.[1] A custom tool, then, would accomplish this task. However, most of you are probably not going to be engaging in that level of statistical or analytical computing, because this specialized task is mainly carried out within the academic research community, and the majority of geography undergraduates are not going into that community after graduation. This is where the second purpose of tools comes in: making ArcGIS Pro easier to use by making it more accessible. The comments below discuss some user interface concepts that can help to understand the value of being able to create custom tools above and beyond adding new functions to ArcGIS.

User interfaces, including ArcGIS Pro, have an inherent tension between two goals: making the interface intuitive versus making the interface expressive. An intuitive interface is easy to use, whereas an expressive interface has a wide range of capabilities. For the most part, an intuitive interface will be limited in expressiveness by only carrying out a few tasks. However, an expressive interface typically gains its expressiveness at the expense of being easy to use. You can probably think of this with ArcGIS easily enough by recognizing that it is quite powerful in the range of things it can do, but it is not user-friendly or intuitive.

At the most extreme end of expressiveness, there are command line interfaces, which is basically scripting. The number of commands is extremely large, and within Python, we can import many packages, which increases the expressiveness even further by importing many more functions, classes, and therefore capabilities. However, as you are learning throughout this book, it is not easy to use.

We can place ArcGIS Pro in an intermediate category. Thanks to a graphical user interface instead of command line, it is easier to use, but there are still many tools and menus and ribbons to navigate. While you would not need to frequently refer to online documentation, as with arcpy, or memorize hundreds of functions and classes, it is still easy to get lost in the various parts of the ArcGIS Pro interface.

At the most intuitive end, within a GIS context, we can perhaps place a tool dialog as an example. If you take just a single tool, you have a clear sense of what the parameters are, what you should do, and how to make the tool run properly. However, as intuitive as a tool with input dialog boxes may be, it only does one task. Therefore, it is as limited in expressiveness as possible. Breaking out of the GIS context, "natural language processing," which is the type of computing for voice assistants like those with the Amazon Alexa and Siri in the Apple iPhone, is perhaps the gold standard of intuitiveness. You can, in theory at least, ask Alexa to set the temperature of your house to a specific setting and (assuming you have a thermostat that

---

[1] ESRI is adding new statistical techniques with, it seems, each new version or ArcGIS Pro, so this statement may soon become outdated. Also, see the incorporation of an R program that does exactly this in Chap. 19.

Supplementary Information: The online version contains supplementary material available at [https://doi.org/10.1007/978-3-031-08498-0_14].

can cooperate with Alexa) have it adjust accordingly.[2] Likewise, you can ask Alexa for the answers to trivia problems, math problems, spelling problems, and so forth. These involve computation, and Alexa is, after all, largely a computer.

Our goal with the task set below, then, is to use a tool to create an intuitive functionality for a commonly executed problem within a building permit process and embed this tool within an ArcGIS Pro project. Lastly, you will share this project and all associated data in a single package so that all the work can be readily shared with less GIS-savvy colleagues.

Task 1: Create a Custom Tool

About the Application

Imagine you work in the local government office that is responsible for evaluating development applications. You are the GIS analyst responsible for maintaining the city's cadaster, which is stored in ArcGIS. A number of your colleagues are responsible for providing applicants with some basic information about their properties and advising them whether their applications are likely to be favorably considered, but they don't have much experience using GIS. So your job is to build a custom tool that will enable your colleagues to use the GIS effectively to answer applicants' questions without requiring them to take a 2-week ESRI training course.

One of the development rules that your local government has is that the size of buildings on a particular block of land cannot exceed 30% of the total land area. So your tool needs to do the following things:

- Do a calculation that figures out what proportion of the land is currently occupied by a structure.
- Tell the user how much of the block is currently occupied by structures.

Start ArcGIS Pro with a new project. When you do this, ArcGIS Pro automatically creates a toolbox for that project. In ArcCatalog, go to that new toolbox through the "Toolboxes" link in the top of the catalog list. Within it should be a toolbox which shares its name with the name of the new project.

Right-click on it and select New→Script. In the Add Script box that comes up, type in the name CalculateArea, and the label "Calculate Area"[3]. Lastly, connect the tool to the script by setting the "Script File" option to the calculateArea.py file that has been provided for you, wherever you have saved it. Click OK to save the edits.

You might have to close and reopen the "Toolboxes" arrow in the Catalog window or otherwise refresh it. ArcGIS won't necessarily update the display to recognize that you have added a tool until you close and reopen it.

What we are going to do is essentially build the dialog box for the tool. In place of the `input` functions that we used in IDLE, we need to have a dialog box in ArcGIS that collects this information from the user. It ends up being a bit more rigorous than the `input` function because it requires us to specify the data type of the input value. It also permits output values. The ability to specify the data type helps us resolve plenty of potentially confusing errors immediately, such as the user entering a feature collection when a raster is expected.

Right-click on the newly created CalculateArea tool and select Properties. At left, choose Parameters instead of General. Now we will enter the two parameters. Click in the first cell underneath Label, and write Block Area. For the Name, put Block_Area or BlockArea (as long as it has no space). We want to restrict this to real numbers by ensuring it is of the type Double. To the right of Block_Area under data type choose Double by clicking on the small ellipsis … that appears and choosing "Double" in the long list that appears when you click on the little triangle at the right end of the bar.

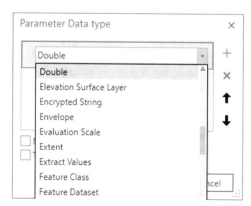

---

[2]See the anecdote about computers not having common sense in Chap. 1 for details on the "in theory" comment.

[3]Observe that the name has no space, as spaces aren't always allowed in parts of ArcGIS file names and file paths. Meanwhile, we are using a label that is more readable.

Do the same in the next row using the display name as BuildingArea, and declare it a double data type. Both parameters should be double type, required, and used as input, as you see below. Click OK.

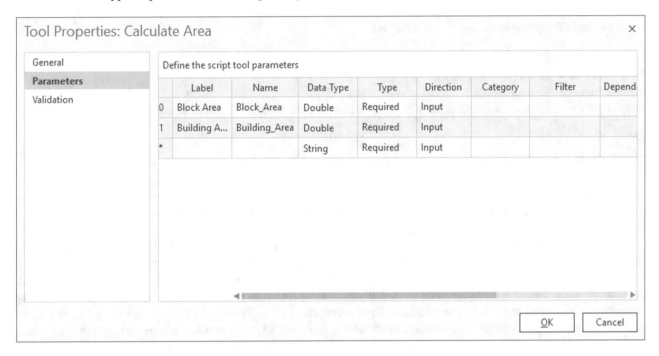

Now if you double-click on the tool name in the Catalog window, it will come up with a tool that looks appropriate for an ArcGIS Pro tool.

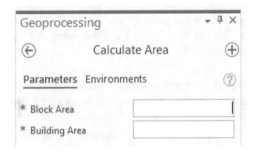

Go ahead and enter some values and click Run. After a (probably very short) while, you'll get a message that "Calculate Area completed."

Click on "View Details," and you'll get, well, a couple more details, but not much. You probably have a couple messages about when it started and when it succeeded, as well as the input parameters you provided. This is all you get because the script does *very* little. It currently consists of one line:

```
import arcpy as ARCPY
```

This might seem odd for this to be the only line in our Python script when it hasn't even been needed except in IDLE, but the automatic importation of the `arcpy` module is only in the Python Window or Notebook. Running a script through a tool in ArcGIS Pro does not, in fact, automatically import this package, even though the tool is within ArcGIS Pro. It is as if you are running the script in IDLE instead.

Now we'll get to actually writing the code. In this chapter and much of what follows, we will be using IDLE to edit our code as we shift across that fuzzy boundary from scripting into programming. To access the Python editor, right click on the calculateArea.py script in Windows File Explorer (wherever you saved it to) and select Edit in IDLE (ArcGIS Pro). It is important to make sure you select the (ArcGIS Pro) option. You'll now see the code editor with the import statement that is the entire script in the main window. This is the code that will run when the button is clicked. In the next task, we will fill out the script to something more useful.

Task 2. Write a Script for a Custom Tool

Below the import, type the following four lines of code:

```
blockArea = ARCPY.GetParameter(0)
buildingArea = ARCPY.GetParameter(1)
percentCover = (buildingArea / blockArea) * 100
ARCPY.AddMessage("Your building covers " + str(percentCover) + " percent of the lot")
```

Question 1: Then, add comments to describe what each line of code is doing.

Save the script. Now test the code by running your tool again in ArcGIS Pro. You shouldn't need to close it or change the values. Just click the Run button again. When you view the details this time, you should have another message, specifically the one that is created in the fourth line above.

By now, you might recognize that running a script within a tool involves some functions which are similar to ones you've seen before in other contexts but which are nonetheless not identical. By having the parameters provided through the dialog box, we need something that replaces the `input` function, and `ARCPY.GetParameter()` provides that service for us. The number in the parentheses[4] is the index of which parameter we want to get from the list. Remember entering in the parameters of block area and building area into the "Parameters" set of the tool properties? This is how we get those, and the index is the position in that list of what we want. Block Area is at the top, so the `blockArea` variable is accessed as parameter #0. Building Area is the second row in this table, so `buildingArea` is created from parameter #1. Likewise, `print` functions will not operate properly in a tool, because there is no output window to print anything to. We solve this through the `ARCPY.AddMessage()` function. We looked at messages in Chap. 9, within the context of a `Result` object. Here we see how this message gets created. We also see how it is possible to use a + operator on strings. It simply appends the second string onto the end of the first, so that "Hello" + "World" becomes "HelloWorld." Note that you have to be careful with spaces if you want things to look normal, so I might suggest "Hello " + "World" instead. In some situations, like the print function, this is equivalent to "Hello", "World", but using comma delimiters creates an error message. Question 2 asks you to identify why this is the case.

Question 2: Using the help documentation of the `AddMessage()` function, or the error message it creates, why should we use the + operator with strings, instead of a comma delimiter like we did with the `print` function?

As said above, an advantage of the tool is that by specifying the type for input variables, a fair number of runtime errors can be removed, but this will not save us from logical errors.

Question 3: What is a logical error that this code might allow a user to make? (including any logical errors you were able to induce while testing the code?) Logical errors would be errors for which the program gives a result the user was not expecting.

One small improvement we could make in the code would be to warn the user if the value for the building area is greater than the area for the block. This might happen if the user is not paying attention to what the input dialog box is prompting them for and enters in the building area first. This is probably more serious than a simple message would warrant, so instead of giving the user a message, we will give them a warning that they may have done something wrong. Add the following lines of code to your program after `ARCPY.AddMessage("...")` line, and save the file:

```
if percentCover > 100:
→ ARCPY.AddWarning("Your building area is greater than your block area. Please check your values
and reenter them.")
```

---

[4]As GetParameter is a function, yes, this number is the parameter of GetParameter. But simply saying "the parameter of GetParameter is the index of the parameter in the parameter list" is confusing.

Try this out with values that should trigger the warning, and you'll see a yellow box that says "Calculate Area completed with warnings." If you proceed to view the details, you'll see the warning, which is displayed in yellow with a triangular yellow warning sign. However, going over to the messages area, both the general message and the warning appear.

Recall from the scenario setup that one of the development rules that your local government has is that the size of buildings on a particular block of land cannot exceed 30% of the total land area. Whereas a ratio exceeding 100% probably indicates a typo, a value between 30% and 100% might indicate an application that should get rejected or at least merit further scrutiny. So for this, we will use the `AddError()` function.

At the end of the script, add these lines to extend the `if` statement to have another `elif` condition:

```
elif percentCover > 30:
→ ARCPY.AddError("Your building might be too big to pass the permit process. Please check your
measurements!")
```

Save the script, and run the tool again with values that should produce this error. Now you'll get a message that the tool failed.

View details to see how it failed, and you'll find that the culprit is the error message you just put into the script. After all, it makes sense that if you add an error to the outputs and run the tool with parameters that should trigger that error message, it would consider the tool to have failed. While we will continue to expand this tool in later chapters, for the time being, we will shift this to sharing the tool for others to use.

Task 3: Sharing the Tool

The last portion of making this useful is to be able to share the tool and associated data with everyone. After all, we want this to be as simple as possible for a novice user of ArcGIS Pro to run it without extensive training. While an initial guess of sharing the .aprx file might make sense, it has implicit instructions to the computer like "get the data for the first layer from that source," and if "that source" is a folder on your computer instead of a common drive, ArcGIS Pro won't be able to find it. Instead, we want to create what ESRI calls a "Project Package" which provides everything we might need.

Load into the map the three shapefiles provided. Then go to the **Share** ribbon and pick the **Project** item, which is the leftmost item in the ribbon and is in the **Package** section. This brings up a window labeled "Package Project" at left, where the tool (probably) had been. Save the package to a file instead of an online account, being sure to choose a name and file location that makes sense. Add a summary of what the project is for, being sure it fits the scenario. Likewise, add some tags. These would be useful if your organization has many projects, and you can use tags to easily find relevant projects for a given task.

Make sure the "Share outside of organization" box is checked, as this should ensure the data is included. Also make sure the "Include Toolboxes" box is checked, because this is necessary to share the toolbox and associated tools. Lastly, if you followed instructions and generated the error message in the previous task, uncheck the "Include History Items" box. Having errors in the geoprocessing history will cause this part to fail if and only if this box is checked.

Click the Package button. If it asks you to save the project first, please do so.

Question 4: The package will be evaluated by the following criteria:
   A. Does it open with the data and tool included?
   B. Does the tool create the appropriate message for building and lot areas that are acceptable?
   C. Does the tool generate the warning if the lot area is bigger than the building?
   D. Does the tool generate the error if the building is too big for the lot?

Task 4: Critical Thinking

In lieu of debugging and unguided programming portions for this chapter, I will instead have you consider the introductory remarks about intuitive versus expressive interfaces, and answer the following question:

Question 5: Keeping in mind the principles of user interaction from the introduction, discuss at least one improvement you can think of to improve on this interface for this specific task of evaluating buildings within lots (if you could theoretically make any type of change—not just ones you know how to implement).

Error Handling

15

Introductory Comments: Strategies for Managing More Complex Code

This chapter picks up where Chap. 14 left off, continuing to improve the usability of ArcGIS Pro within the context of the cadastral application from the previous lab. There are two ways we will try to improve things. The first will be to change the parameters to allowing the user to either enter the lot number or enter the block and section number for the building, instead of having to know the size of the lot and the size of the building footprint. The second improvement will be to improve how errors get handled, so the user is less likely to be the recipient of hard-to-read error messages.

To make these improvements, we will need to have a more complex structure to our Python code, including defining our own functions and using more than one function. Most of the concepts that underlie this added complexity are not new. The first section of the book provided almost everything that you need to know, whether that is defining a function and calling it or using a conditional statement. The only new material is some additional means of managing and creating error messages, which will be introduced as it is entered into the program.

The bigger changes are, in a way, the length of the program, as the file gets longer. As it does so, and especially in later chapters as the complexity continues to increase, you might find it useful to use the strategies introduced earlier of pseudo-code outlines and/or flow charts to better plan the code, as well as understand what it is doing. Lastly, this chapter illustrates a realistic aspect of the evolution of code within a program. The additions and changes do not take place in a nice, neat, and linear fashion but instead are added to different portions of the code out of order. This mirrors what often happens in practice, especially for large projects that are too lengthy to be developed sequentially from start to finish in one sitting.

Task 1: Update the Parameter List for the Tool

The first part of making these upgrades is figuring out what information we need from the user to make this updated approach possible. We will change the set of parameters our tool accepts to accomplish this. First, we need to know which set of information they are giving. This means changing the first parameter from the lot area to one that sets the option they are taking: lot number or building block/section number. Change the Label to "Option," and the Name to "Option." We want this to be a String, so change the Data Type to String. There will be a new setting in the parameter list that we will use, because we only want one of two options to be selected. There is a setting called "Filter." Select it, and pick "Value List." Type in the first option ("Lot Number"—but without the quotes), and click the + button to add another row. Then type in "Building Block & Section." It is critical to have these EXACTLY as they appear here. Typos in these strings will cause serious issues with the code.[1] It should appear as you see below.

---

[1] Unless you have matching typos in the code, which is not a recommended strategy.

© The Author(s), under exclusive license to Springer Nature Switzerland AG 2022
J. Conley, *A Geographer's Guide to Computing Fundamentals*, Springer Textbooks in Earth Sciences, Geography and Environment, https://doi.org/10.1007/978-3-031-08498-0_15

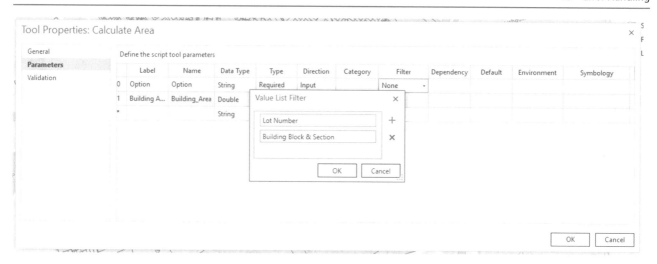

Click OK. Then change the second parameter "Label" to Lot Number, and set its data type to string.[2] We also want it to be an Optional parameter. Likewise, we need to add two more optional parameters for the building block number and building section number. These can have the data type of Long. When you have completed it, the parameter list should look like this.

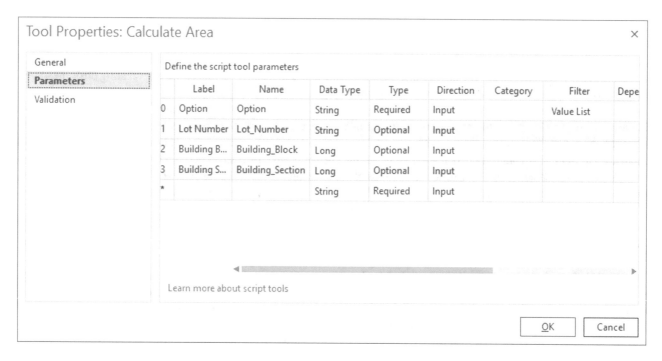

Question 1: Why did we set the last three parameters to be optional, instead of required?

Task 2: Create Functions to Make the Code More Manageable

Now that we have the parameter list changed to accommodate these upgrades, we need to change the script. As mentioned in the introductory comments, flow charts might be helpful. To begin with, let's look at a flow chart of the existing program for our tool (Fig. 15.1).

---

[2]Yes, string, even though it is a number. It turns out that the lot numbers are, in fact, so numerically large that they do not even fit within a Long integer data type. Therefore, we will have to treat them as a string, which is acceptable because it is a categorical variable anyway.

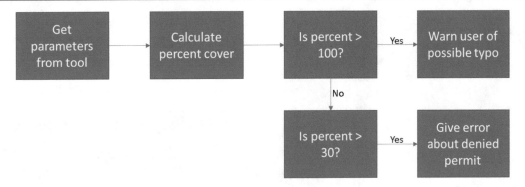

Fig. 15.1 Beginning flow chart

First, we will change the section of the code that collects the parameters to get all four parameters instead of just the ones from before, which by itself does not require any change to the flow chart. It does, though, mean that we will need more code, because we no longer have `buildingArea` and `blockArea` directly from the user, so will have to look them up from the identifiers that the user does provide:

```
import arcpy as ARCPY


option = ARCPY.GetParameter(0)
lotNumber = ARCPY.GetParameter(1)
buildingBlock = ARCPY.GetParameter(2)
buildingSection = ARCPY.GetParameter(3)
```

Because the user can provide the identifiers in one of two options, it makes sense that after getting the parameters, we would now need a conditional statement to identify which option is being selected. Our script will have several ways to be completed, because of the two options for the first parameter. This directly increases the complexity but also indirectly increases the complexity. In both options, we will want to do the calculations that are the remainder of the existing flow chart. However, if we think we might be using an if/else statement to handle the two options, why have the same calculation code repeated in both parts of the if/else statement? To handle this, using a function is commonplace. We will first put the main logic of the code into a function. This gives the next flow chart, which is obviously incomplete now that not all parts are even connected (Fig. 15.2).

This entails a few changes in the code. Recall that functions must begin with a line that starts with the `def` keyword, as this is what defines a function. We would need a couple parameters to carry out these calculations: building area and block area. These are set as parameters into the `def` statement. Once that is done, the only change for now is to make sure we use the `return` keyword to return the calculated value. We will get to the conditional statement later:

```
def calculate(buildingArea, blockArea):
    percentCover = (buildingArea / blockArea) * 100
    if percentCover > 100:
        ARCPY.AddWarning("Your building area is greater than your block area. Please check your
values and reenter them")
    elif percentCover > 30:
        ARCPY.AddError("Your building might be too big to pass the permit process. Please check
your measurements!")
    return percentCover
```

Working backward from the newly defined function, we now need to have ways of taking the information that the user provided and using it to find the building area and block area. Once that is done, those value can be the inputs into this function. Because of how we are setting up the tool, separate functions for the lot option and the building option would be helpful. The `def` statements are shown below, but do not yet contain any code. That will be done the next task:

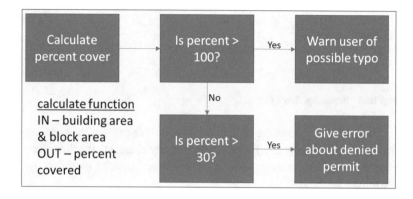

Fig. 15.2 Introducing a function

```
def calculateFromLot(lotNum):

def calculateFromBuilding(bBlock, bSection):
```

Task 3: Adding a Try/Except Structure

This task starts to handle errors in a manner that we can use to make the tool easier to use. In this context, errors are things that you have encountered many times while debugging, such as a `SyntaxError` or `TypeError`. We can do better than waiting for Python to generate an error for us. It might initially seem redundant to create an error when we already have the `ARCPY.AddError()` function that can provide an error message to the user. However, using Error objects gives us more flexibility in handling them. If we know something isn't right, such as a building being larger than its lot, we can create the error ourselves. Also, if a situation arises that creates an error, but we can provide a workaround, that, too, is something we might want to address. In this manner, we can use errors more intelligently to make life easier on the user.

Errors are created through the `raise` statement. These send error messages. Those messages then get received within the code that called the line that generated the error. If that code cannot handle the error, it sends the error message back out and back up the parsing stack. If you think back to the Traceback in Chap. 6 on debugging, where the error was found in line 23 of a function but was when that function got called on line 68. The error is raised in the function on line 23, and the error message is received on line 68. In that chapter, line 68 had no way of dealing with it, so the script stopped running, and you, the programmer, were given the error message. Changing the function to raise errors changes the flow chart slightly, from giving warnings and errors to raising `ValueError` objects. This means that a value is wrong, as would be the case if the building area were greater than the lot area, or if the percent of the lot which is covered could lead to a denied permit (Fig. 15.3).

See below for how these were changed, including having a message that is more useful to the user inside the parentheses. This is another useful purpose of having custom errors; you can control what the messages are and therefore can try to make them as useful as possible to the users:

```
def calculate(buildingArea, blockArea):
    percentCover = (buildingArea / blockArea) * 100
    if percentCover > 100:
        raise ValueError("Your building area is greater than your block area. Please check your
values and reenter them")
    elif percentCover > 30:
```

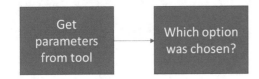

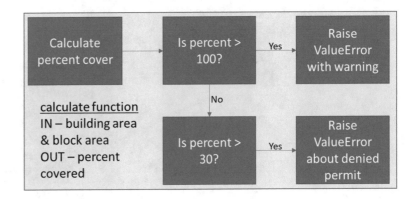

Fig. 15.3 Changing the messages to errors

```
    raise ValueError("Your building might be too big to pass the permit process. Please check
your measurements!")
    return percentCover
```

Next, we will work on the other part of error handling, catching an error to prevent the user from seeing the error message. This involves a new complex statement type: a `try/except` block. Python will try to run the code in the `try` block. If it works, great. The code inside the `try` block is fully executed and the `except` part is ignored. If there was an error, however, Python will see if the type of error raised is in the `except` section of the structure. If that type is listed, then the code inside the relevant `except` block is executed. If that type of error is not listed, then the error that occurred gets passed on to ArcGIS Pro and almost certainly results in a bunch of error messages and red text where you expected to get your output. Lastly, notice that we except the `ValueError` as a specific variable called `ve`. This allows us to access the message that is inside the error, which was put there when we created the message. The `ValueError` is explicitly used here because it is what is raised within the function. Put this at the very end of the program file:

```
try:


except ValueError as ve:
 ARCPY.AddError("There was an error during the operation of the tool. " + str(ve))
```

Now we will fill in code for the `try` block, which largely entails moving the parameter acquisition lines from the start of the script, and finally adding that `if/elif` statement to control what happens based upon the value the user chose within the option parameter. Let's first plan it out by updating our flow chart (Fig. 15.4).

To update the code, it helps to recognize that the options are restricted to the values in the Value List Filter from setting up the new dialog box. We will be able to use these restricted values in the conditional statement, knowing that they must match, as long as there are no typos. This fills in the `try` block as below:

```
try:
    option = ARCPY.GetParameter(0)
    if option == "Lot Number":
        lotNumber = ARCPY.GetParameter(1)
        pctCover = calculateFromLot(lotNumber)
    elif option == "Building Block & Section":
```

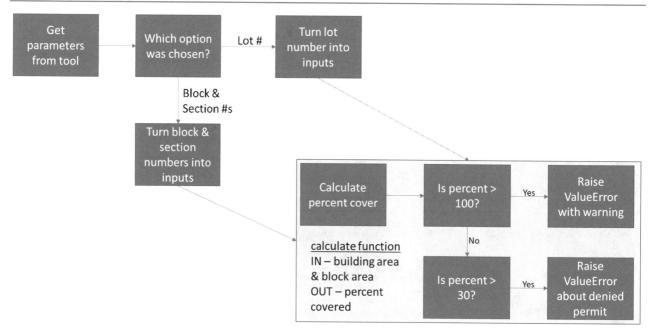

Fig. 15.4 Adding the option calculations to the flow chart

```
        buildingBlock = ARCPY.GetParameter(2)
        buildingSection = ARCPY.GetParameter(3)
        pctCover = calculateFromBuilding(buildingBlock, buildingSection)
    ARCPY.AddMessage("Your building covers " + str(pctCover) + " percent of your lot. It should
pass.")
except ValueError as ve:
    ARCPY.AddError("There was an error during the operation of the tool. " + str(ve))
```

Question 2: Why can we say with a fair amount of confidence that if the code reaches the line with the ARCPY. AddMessage function, that the values will indeed pass.[3]

Task 4: We Will Now Fill in the Two Intermediate Functions

The last things to fill out are the two remaining intermediate functions, converting the user inputs into what we need for the `calculate` function as it is defined. Once these are filled out, we will have three functions and a try/except block of code. While you might think this transformation from a lot number to the building and lot areas is simple at first, the flow chart for one subtask might disabuse you of the notion. You should be able to recognize most, if not all, of what is going on in the flow charts and functions as I have them below. A query is set to search for the relevant information, whichever approach the user takes. Once this is done, cursors are used to get the relevant information from the table. It is assumed, in the first function, that there is only one lot that has each lot number. This assumption from Question 3, however, is not present in the building cursor. Because a lot can have multiple buildings, we must cycle through all the buildings, looking for any of them which have the right block and section values. Also notice that I have the call to the calculate function directly in the return statement, illustrating the ability to call one function from within another function.

Question 3 (10 points): How does this assumption justify the use of the `.next()` function?

---

[3]Assuming, for the time being, that the calculateFromLot and calculateFromBuilding functions are going to be implemented correctly.

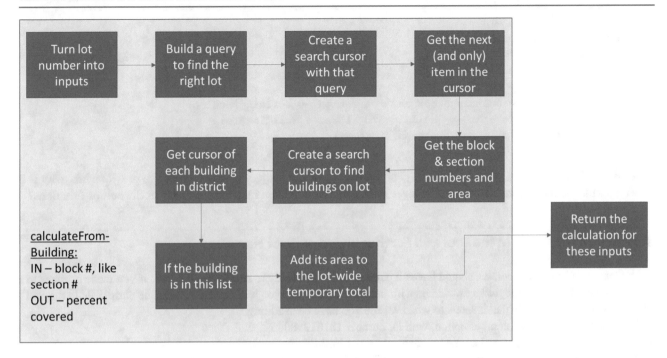

In the code for this is presented below:

```
def calculateFromLot(lotNum):
    query = '"KEY" = ' + str(lotNum)
    search = ARCPY.da.SearchCursor('lots', ["AREA", "KEY", "BLOCK", "SECTION"], query)
    lot = search.next()
    lotArea = lot[0]
    lotBlock = lot[2]
    lotSection = lot[3]
    buildings = ARCPY.da.SearchCursor('buildings', ["BLOCK", "SECTION", "F_AREA"])
    buildingsTotalArea = 0
    for building in buildings:
        buildingBlock = building[0]
        buildingSection = building[1]
        buildingArea = building[2]
        if lotBlock == buildingBlock and lotSection == buildingSection:
            buildingsTotalArea = buildingsTotalArea + buildingArea
    return calculate(buildingsTotalArea, lotArea)
```

This assumption that there is one ID number per lot, however, is not present in the building cursor. Because a lot can have multiple buildings, we must cycle through all the buildings, looking for any of them which have the right block and section values, and keep a running total of all buildings, which gets involved in more of the calculations. Also, noticing that the lot file has the same information, we can reuse this query for the lot file and again use the next function, as there is only one lot with that ID number:

```
def calculateFromBuilding(bBlock, bSection):
    query = '"BLOCK" = ' + str(bBlock) + ' AND "SECTION" = ' + str(bSection)
    buildings = ARCPY.da.SearchCursor('buildings', ["BLOCK", "SECTION", "F_AREA"], query)
    lot = ARCPY.da.SearchCursor('lots', ["AREA", "BLOCK", "SECTION"], query).next()
    lotArea = lot[0]
    lotBlock = lot[1]
    lotSection = lot[2]
```

```
buildingsTotalArea = 0
for building in buildings:
    buildingBlock = building[0]
    buildingSection = building[1]
    buildingArea = building[2]
    if lotBlock == buildingBlock and lotSection == buildingSection:
        buildingsTotalArea = buildingsTotalArea + buildingArea
return calculate(buildingsTotalArea, lotArea)
```

Question 4: Recalling that you are (probably) still working in the project package from the previous lab, why is it acceptable—*in this instance*—to use the layer names of `lots` and `buildings` directly in the cursor functions?

Task 5: Run the Script in the Tool and Further Improve Error Handling

With these functions filled out, we should have all parts of our script ready to run within the tool. If you made changes to a script that isn't in an obscure folder associated with the project package, you will need to change the Script File in the tool's properties→General area, so that it refers to whatever file you have been editing.[4]

Run the tool with the lot number option, with lot number 12181240006.

Question 5: What is the percent of the lot covered by the building in this lot?

However, entering a lot number that doesn't exist gives a "`StopIteration`" exception[5]. This is something that is going to be confusing to a novice user, so we want to fix this so that the user at least gets an easier to understand error message. We can have multiple except clauses in a `try`/`except` block, just like we can have multiple `elif` clauses. In this case, a `StopIteration` exception is caught, and its message gets replaced with a betting response through the `ARCPY.AddError()` function. The full try/except block is presented below, with the new code in the last three lines:

```
try:
    option = ARCPY.GetParameter(0)
    if option == "Lot Number":
        lotNumber = ARCPY.GetParameter(1)
        pctCover = calculateFromLot(lotNumber)
    elif option == "Building Block & Section":
        buildingBlock = ARCPY.GetParameter(2)
        buildingSection = ARCPY.GetParameter(3)
        pctCover = calculateFromBuilding(buildingBlock, buildingSection)
    ARCPY.AddMessage("Your building covers " + str(pctCover) + " percent of your lot. It should
pass.")
except ValueError as ve:
    ARCPY.AddError("There was an error during the operation of the tool. " + str(ve))
except StopIteration as si:
    if option == "Lot Number":
        ARCPY.AddError("Your lot number doesn't exist.")
```

To see what is happening, let's follow the exception as it gets raised and propagated through the program here. The `StopIteration` exception takes place when the `next()` function is called on the cursor named search in `calculate-FromLot()`. That exception gets raised there, and execution of the `calculateFromLot()` function stops immediately. Then the exception goes up to what called that function, which is in the fifth line of the try/except block. Since this is in a

---

[4]Within a project package, at least, this is an easy mistake to make, as the package copies the file to a new project in a fairly obscure area, so changing the script to where you have the previous lab materials saved is probably easier to work with.

[5]There is a difference between exceptions and errors, although the difference is beyond the scope of this book.

try/except block, the computer looks to see if there is an `except StopIteration` clause. Previously, there wasn't so execution of the entire program stopped, and you got an error message in the tool. Now there is an `except StopIteration` clause, so what happens in that except block of code is where the program goes. In this case, it still provides an error message to ArcGIS Pro, although we are able to give an error message that is far more useful than "StopIteration."

Entering a nonexistent combination of block and section number here doesn't generate an error. It just sits there and might eventually report an answer of zero. This means that for us to treat it similarly, we need to raise the error ourselves. This can be done by counting the number of buildings in this block/section combination and raising our own `StopIteration` if it has the value of zero.[6] The `calculateFromBuilding` function is presented below, with the new code being that which involves the variable `buildingCount`:

```
def calculateFromBuilding(bBlock, bSection):
    query = '"BLOCK" = ' + str(bBlock) + ' AND "SECTION" = ' + str(bSection)
    buildings = ARCPY.da.SearchCursor('Narra_foot', ["BLOCK", "SECTION", "F_AREA"], query)
    lot = ARCPY.da.SearchCursor('Narra_block', ["AREA", "BLOCK", "SECTION"], query).next()
    lotArea = lot[0]
    lotBlock = lot[1]
    lotSection = lot[2]
    buildingsTotalArea = 0
    buildingCount = 0
    for building in buildings:
        buildingCount = buildingCount + 1
        buildingBlock = building[0]
        buildingSection = building[1]
        buildingArea = building[2]
        if lotBlock == buildingBlock and lotSection == buildingSection:
            buildingsTotalArea = buildingsTotalArea + buildingArea
    if buildingCount == 0:
        raise StopIteration
    return calculate(buildingsTotalArea, lotArea)
```

The `try/except` statement can then get an `else` clause in the except `StopIteration` lines. They are the last lines of the newly edited `try/except` block here:

```
try:
    option = ARCPY.GetParameter(0)
    if option == "Lot Number":
        lotNumber = ARCPY.GetParameter(1)
        pctCover = calculateFromLot(lotNumber)
    elif option == "Building Block & Section":
        buildingBlock = ARCPY.GetParameter(2)
        buildingSection = ARCPY.GetParameter(3)
        pctCover = calculateFromBuilding(buildingBlock, buildingSection)
    ARCPY.AddMessage("Your building covers " + str(pctCover) + " percent of your lot. It should
pass.")
except ValueError as ve:
    ARCPY.AddError("There was an error during the operation of the tool. " + str(ve))
except StopIteration as si:
```

---

[6]Using a StopIteration here when the problem has nothing to do with anything that is iterating, like a for loop, is officially not best practice. However, using the same kind of error as the other option streamlines our try/except code.

```
    if option == "Lot Number":
        ARCPY.AddError("Your lot number doesn't exist.")
 else:
        ARCPY.AddError("Your building block/section combination doesn't exist.")
```

Likewise, a lot without any buildings isn't necessarily indicative of a correct data entry, so it is something we might want to warn user about. This is done through another `ValueError` that gets raised toward the end of the `calculateFrom-Lot` function:

```
def calculateFromLot(lotNum):
    query = '"KEY" = ' + str(lotNum)
    search = ARCPY.da.SearchCursor('Narra_block', ["AREA", "KEY", "BLOCK", "SECTION"], query)
    lot = search.next()
    lotArea = lot[0]
    lotBlock = lot[2]
    lotSection = lot[3]
    buildings = ARCPY.da.SearchCursor('Narra_foot', ["BLOCK", "SECTION", "F_AREA"])
    buildingsTotalArea = 0
    buildingCount = 0
    for building in buildings:
        buildingBlock = building[0]
        buildingSection = building[1]
        buildingArea = building[2]
        if lotBlock == buildingBlock and lotSection == buildingSection:
            buildingCount = buildingCount + 1
            buildingsTotalArea = buildingsTotalArea + buildingArea
    if buildingCount == 0:
        raise ValueError("Your lot appears to contain zero buildings. Please confirm that this is
the right lot.")
    return calculate(buildingsTotalArea, lotArea)
```

The last thing we will do is check the situation of the user having the wrong option selected, so that they selected the lot number option but entered the block and section number or selected the building-based option but entered a lot number. Leaving things blank gives different errors for the different options, but we can check both of them with similar logic. These checks are done at the beginning of the two calculateFrom... functions. The two checks for emptiness are different because the parameters are different data types, where an empty box in the tool dialog is turned into an empty string for a string data type parameter, but an empty box in the tool dialog is turned into a 0 for a numerical data type parameter:

```
def calculateFromLot(lotNum):
    if lotNum == '':
        raise ValueError("Your lot number parameter is empty. Please fill it in or change the option.")
    query = '"KEY" = ' + str(lotNum)
    search = ARCPY.da.SearchCursor('Narra_block', ["AREA", "KEY", "BLOCK", "SECTION"], query)
    lot = search.next()
    lotArea = lot[0]
    lotBlock = lot[2]
    lotSection = lot[3]
    buildings = ARCPY.da.SearchCursor('Narra_foot', ["BLOCK", "SECTION", "F_AREA"])
    buildingsTotalArea = 0
    buildingCount = 0
    for building in buildings:
        buildingBlock = building[0]
        buildingSection = building[1]
```

```
            buildingArea = building[2]
            if lotBlock == buildingBlock and lotSection == buildingSection:
                buildingCount = buildingCount + 1
                buildingsTotalArea = buildingsTotalArea + buildingArea
        if buildingCount == 0:
            raise ValueError("Your lot appears to contain zero buildings. Please confirm that this is
the right lot.")
        return calculate(buildingsTotalArea, lotArea)


def calculateFromBuilding(bBlock, bSection):
    if bBlock == 0 or bSection == 0:
        raise ValueError("Your building block and/or building section is empty (or zero). Please
fill it in or change the option.")
    query = '"BLOCK" = ' + str(bBlock) + ' AND "SECTION" = ' + str(bSection)
    buildings = ARCPY.da.SearchCursor('Narra_foot', ["BLOCK", "SECTION", "F_AREA"], query)
    lot = ARCPY.da.SearchCursor('Narra_block', ["AREA", "BLOCK", "SECTION"], query).next()
    lotArea = lot[0]
    lotBlock = lot[1]
    lotSection = lot[2]
    buildingsTotalArea = 0
    buildingCount = 0
    for building in buildings:
        buildingCount = buildingCount + 1
        buildingBlock = building[0]
        buildingSection = building[1]
        buildingArea = building[2]
        if lotBlock == buildingBlock and lotSection == buildingSection:
            buildingsTotalArea = buildingsTotalArea + buildingArea
    if buildingCount == 0:
        raise StopIteration
    return calculate(buildingsTotalArea, lotArea)
```

At this point, you should have the full script entered and working within the tool.

Task 6: Debugging Exercise

The debugging exercise for this chapter is unrelated to the rest of the chapter with respect to the application to avoid introducing any errors or problems into the main program we are working with. It is, instead, a self-contained piece of code with a try/except statement that can be entered and debugged within a single cell of the Python Notebook. When fixed, it should print out a set of messages with respect to the numbers in the list and the criteria set forth in this chapter regarding the permitted size of buildings with respect to their lots. There are four errors. Make sure the requirements are met, as logical errors could lead to raising a ValueError that should not be happening:

```
lst = [25, 36, 135, -3, 25, 14, 17, 504, 42]
for item in lst:
    try
        print("The building covers " + str(item) + "% of the lot.")
        if item < 0:
            raise ValueError("Negative building size? Is this a typo?")
        elif item > 100:
            raise ValueError("Building larger than lot? Is this a typo?")
        elif item < 30:
            raise ValueError("Building is too big for the lot! Revise your plans to meet the
regulations.")
```

```
        else:
            print("Permit approved.")
    except ValueError:
        print("There is a problem with the application: " + ve)
```

Task 7: Unguided Work

We used the fact that the block number and section number are in the lot file when extracting the lot size within the `calculateFromBuilding` function. However, as set up right now, our tool will not allow the user to use the lot-based option and enter the block and section numbers instead of the lot number. For example, the user should be able to have the Option be "Lot Number," but leave the Lot Number itself blank, and enter Block #23 and Section #41, and still have it work.

Make changes **to the script** to allow this to happen. The only changes you may make to the tool is altering the label for the lot-based option, as it isn't using a lot number any more.[7] This, however, is not required:

The tool will be evaluated based upon the following expected behaviors:

The tool should give a proper response for a lot that contains buildings.

The tool should still give an error if the combination of block and section number does not exist.

The tool should still give a warning if the combination appears to contain no buildings, but the lot exists.

Note: There are multiple ways of accomplishing this. As long as the script behaves correctly regardless of what values I enter, you will get credit regardless of your approach.

---

[7] Translation. Don't just tell them to use the block/section option for either file. Allow the option parameter to be "Lot Number" (or "Lot Information" should you choose to edit it), the lot number parameter to be blank, and the block and section number information to be filled out.

Creating Custom Classes

Introductory Remarks: Strategies for Creating and Managing Classes

Once again, this chapter continues the progression of making the usability of ArcGIS Pro better within the context of approving or denying building permits for a city. This requires changes, again to the tool, to align it better to the purpose of having a tool. In this case, we will change it to carry out more of the approval process. Specifically, we want to be able to enter a lot ID number (as was done in parameter #1 by the end of the previous lab), the area of a proposed building,[1] and the type of building (residential, commercial, etc.). Given this information, the tool will report the estimated appraised tax bill or will give an ArcGIS error message about why the proposed building is unlikely to be approved.

While carrying out these changes will make life easier for the user, it will make the code still more complex. To help evaluate the proposed buildings, we will create two new classes. One is called `Parcel`, which will represent the combination of the lot and building(s) contained on that lot. The other is called `ApprovalError`, which is an extension of the `ValueError` class to represent a specific version of value error which violates the guidelines of the approval process.

Before we go further, we will revisit ideas about classes which were introduced in the ArcPy chapter. The object-based approach allows us to have variables which are of more complicated types than the standard suite of text, numbers, and Boolean values. Each object has a blueprint of its class, which represents the set of things it contains. The main parts to this blueprint are properties, which are akin to variables specific to this object, and methods, which are functions that operate on this object. Recall this diagram from the ArcPy chapter, showing some of the properties on the left and methods on the right associated with the class `ArcGISProject`, and we have the variable `aprx` which is an object of the type `ArcGISProject` (Fig. 16.1).

Just as we would want to use custom functions to be able to make code more reusable and readable, if we have a group of variables and functions which are related to each other, and which we may find ourselves calling repeatedly, perhaps it would be wise to group them into a single class with those variables and functions converted into properties and methods. In the scenario here, while we are receiving applications to create a building, the evaluation is done at the scale of a parcel. Since the information for the parcel is split between the file for the buildings and the file for the lots, we may want to create a class that combines the relevant information from the two different files into a single object. Once that information is combined, we can then continue the analysis with the parcel holding all of our information in a single place for us.

Recall from the data description that we can have classes which extend other classes. Just as we had a `FeatureClass` which extended `Table`, and implemented all the properties and methods associated with the `Table` and adds additional properties and methods, we can create classes which extend existing classes. These new classes can then make use of the existing class' properties and methods, add new properties and methods, as well as override existing methods. Overriding a method keeps the method in existence, but changes how it gets carried out. For example, highlighting the selected items in a table in ArcGIS Pro will make those rows appear differently within the attribute table. Highlighting selected items is also carried out by a feature class, but the way they get highlighted is different. The relevant rows of the attribute table still appear differently, but the features in the map also get highlighted, so we can see how the action of highlighting items is carried out in a different manner. Within Python, overriding methods is especially notable for what are called "dunder" methods. These

---

[1] Ideally, this would accept a polygon, from which area can be calculated, along with comparing it to the polygon of the lot to evaluate setbacks, but while polygon is a valid input parameter type, giving the tool the ability to input a polygon would probably be counterproductive to the idea of trying to make this easier for the user, not harder.

© The Author(s), under exclusive license to Springer Nature Switzerland AG 2022
J. Conley, *A Geographer's Guide to Computing Fundamentals*, Springer Textbooks in Earth Sciences,
Geography and Environment, https://doi.org/10.1007/978-3-031-08498-0_16

```
┌─────────────────────────────────────────┐
│            ArcGISProject                 │
├─────────────────────────────────────────┤
│ activeMap: Map                           │
│ dateSaved: DateTime                      │
│ defaultGDB: String                       │
│ defaultToolbox: String                   │
│ documentVersion: String                  │
│ filePath: String                         │
│ homeFolder: String                       │
├─────────────────────────────────────────┤
│ save() : void                            │
│ listLayouts({wildcard : String}) : List  │
└─────────────────────────────────────────┘
```

Fig. 16.1 Class diagram of a class with properties and methods

are built-in methods that apply to most (if not all) classes, and they get their name because the method names always start and end with two underscores (a double underscore or "dunder"). The two that we will use in this chapter are the `__init__` and `__str__` methods. The first of these is the class' constructor method, which is the code that is run when an object of this class is created and is typically where the properties of this new object are set. Without it being made explicit, you have used `__init__` methods repeatedly. Any time you use a class' constructor method, you implicitly use the `__init__` constructor. Therefore, when you create a `Describe` object through calling `arcpy.Describe()`, this is, behind the scenes, calling the `__init__` method of the `Describe` class. The second of these is how the object will be represented when it is turned into a string. It isn't an exact match for this, but you can probably think of it as what happens to an object when you call the `str()` function on that object.

When programming within a class, you will need to be able to refer to the object of that class which is the instance of the class that all the properties and methods are acting upon. For example, when you call the `aprx.listLayouts()` method through the `aprx` variable in the figure above, the code within the `listLayouts` method needs to be able to access the `aprx` variable. This is not as straightforward as you might initially think, because the code has to be written such that it doesn't matter what the variable is called. We don't know if the `ArcGISProject` object is called `aprx`, `myProject`, `currentProj`, or even `iHateArc`. For this reason, it is accessed within the code through a special variable which is usually called `self`.[2] When creating the variable in the `__init__` method, `self` is the first variable. Likewise, when passing the variable through methods, the first parameter is always `self`. Just as you would use `aprx.listLayouts()` in the Python Window, you would use `self.listLayouts()` if you needed to access the list of layouts while in another method of the `ArcGISProject` class. Also, if you need to access a property within a method, like the `save()` method needing to be able to access the `filePath` property, the code within `save()` would use `self.filePath` to get to that property.

As we update this tool to provide better functionality and create our two classes: `Parcel` and `ApprovalError`, we will see these concepts in action.

Task 1: Update the Tool Dialog Box

As stated above, we want to be able to enter three input parameters: a lot ID number, represented as a string; the area of a proposed building, represented as a Double; and the type of building, representing a string restricted to be from the following list of values: COMMERCIAL, COMMUNITY, EDUCATION, GOVERNMENT, HEALTH, RECREATION/TOURIST, or RESIDENTIAL. These are taken directly from the type field within the file of buildings, so must be exactly as they appear, including in all caps. Otherwise, the code will not work.

Question 1: Please make the changes to the tool to make it appear as below. Note the changes to the display name of the tool ("Propose New Building"), the three parameters, the display text for them, and the ability to select from the list of building types. Not shown: all three parameters are now required. This should be a review of skills from the previous two chapters.

---

[2]It is not required to be called self, but standard practice is that it almost always called this. As such, I'll just refer to it as self from here on out.

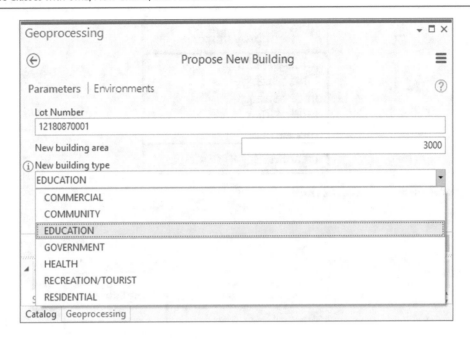

Task 2: Plan Out the Classes with UML, Flow Charts, and Pseudocode

Before we start writing out the code, we will want to plan what goes here. Without having a plan of attack, we will probably be staring at a blank screen not knowing where or how to begin. The first step is to identify what we need to include as the properties and methods for our two new classes. The simpler one is the `ApprovalError` class. It has, as shown below, some information that is going to be useful for creating the error message, specifically the type of error, the lot ID, two relevant pieces of information in generically titled properties of `info1` and `info2`,[3] and the error message we will want to give to the user. Lastly, notice that for the purposes of this exercise, instead of a variable name, like `aprx` above, the object is called `self` below the box (Fig. 16.2).

Using the information in the properties, we then want to plan out the `createMessage()` method. A flow chart for this is below, assuming error types follow these codes: 1 is a proposal denied because the building was too large, and 2 represents a proposal denied because the proposed building is of a type that does not match buildings already on the lot (Fig. 16.3).

The parcel class is more complex than the `ApprovalError` but in terms of what it needs as the blueprint of methods and properties is not much more so. It also has five properties, but it has two methods instead of one: an `assess()` method which is specific to calculating the value and the primary method called `propose()`, which takes in the area and type of the proposed building and decides whether to approve the proposal or not (Fig. 16.4).

Of the two methods, assess is by far the simpler one. It just applies a formula to the relevant information and returns the result of that calculation (Fig. 16.5).

The only potential hang-up you might notice is that we are using the building type, which is one of the properties of the class, to find a tax rate. It so happens that we can have variables within the object which are, in effect, private instead of public.[4] One such variable allows us to get a tax rate for the building based upon its type.

Question 2 (5 points): Knowing what you do about Python, what data type would you use for this variable linking tax rates with building types? Why?

The other method carries out the calculations for the approval process, so it is more complex. We need to check the two conditions for approving or denying building proposals: whether the new building matches the type of existing buildings, if

---

[3] The two pieces of information here are different for the different error types, so a more specific name cannot be given. This suggests that maybe we would want to have different extensions of the ApprovalError class, but that seems a bit too much for this exercise. It also illustrates that, while it is presented here in a neat and tidy fashion, there is some iteration in the process of deciding what is or isn't needed, as I didn't realize I would need two generic pieces of information to create the error messages until I started working on the createMessage() planning below.

[4] Some programming languages allow you to explicitly make properties and even methods private. Python is not one of those, to my knowledge, so we are using properties whose existence we would not advertise, so we can treat them as private.

ApprovalError

errorType : String
info1 : String
info2 : String
lotID : String
message : String
createMessage() : String

Fig. 16.2 Crate diagram of new `ApprovalError` class

Fig. 16.3 Flow chart for self.
createMessage() in
ApprovalError

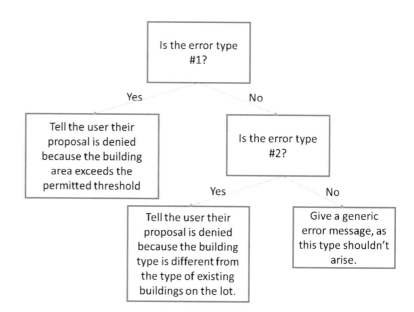

Parcel

buildingCount : Integer
buildingType : String
totalBuildingArea : Double
lotID : String
lotArea : Double
assess() : Double
propose(newArea : Double, newType : String) : Double

Fig. 16.4 Class diagram of new `Parcel` class

Fig. 16.5 Flow chart for
`self.assess()` method of
new `Parcel` class

Get the tax rate for buildings based upon its type	→	Calculate the total estimated tax using this formula: base rate * lot area + building rate * building area	→	Return the value of this calculation

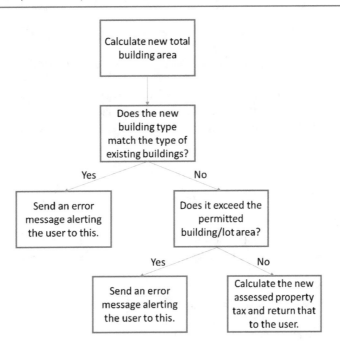

Fig. 16.6 Flow chart for `self.propose()` method of `Parcel` class

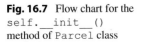

Fig. 16.7 Flow chart for the `self.__init__()` method of `Parcel` class

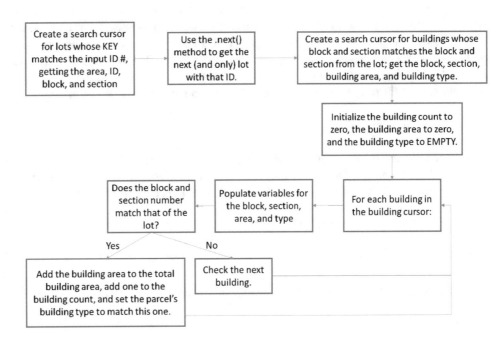

there are any, and whether the new building will push the ratio of total building area to lot size above a predetermined thresh-old. If it passes these tests, the new assessed value can be returned to the user, while if it doesn't, we can raise an `ApprovalError`, as defined already (Fig. 16.6).

The more observant among you might have a few remaining questions. Specifically, there has to be something that happens in between giving a new parcel a lot number, the size and type of a proposed building, and the evaluation that happens in the `propose()` method. After all, the `propose()` method uses information like the total building area, the type of existing buildings, and the lot size, none of which are provided to the tool directly. We still need to get this information somehow. If you recall, though, the previous lab on error handling, we were able to find out information like this just from the lot ID number. In fact, much of this code will be taken from the `calculateFromLot()` function that you wrote in that chapter (Fig. 16.7).

The question becomes where do we put this code? As it is something that happens when we initially create the parcel, it belongs in the `__init__()` dunder method.

Task 3: Create Shells of the Classes

Now that we have our plans, we can start to fill in the code for these classes. While we are at it, we can change the code for the try/except block at the end of the file. It will, in fact get much simpler, as we will see. We just need to get the parameters from the tool, create our parcel, run the propose() method, and give the user a message. If there are any problems, we will receive an ApprovalError, which is caught in an except clause. We also catch a StopIteration error, as before, to improve the user friendliness for the situation where the user enters a lot number which doesn't exist.

```
try:
 lotID = ARCPY.GetParameter(0)
 newBuildingSize = ARCPY.GetParameter(1)
 newBuildingType = ARCPY.GetParameter(2)
 permitParcel = Parcel(lotID)
 newAppraisal = permitParcel.propose(newBuildingSize, newBuildingType)
 ARCPY.AddMessage("Your permit should be approved, and the new tax amount is estimated to be " +
str(newAppraisal))
except ApprovalError as ae:
 ARCPY.AddError("There was an error during the approval process. " + str(ae))
except StopIteration as si:
 ARCPY.AddError("Your lot number doesn't exist.")
```

Defining our new classes introduces another keyword, which is like def, but for classes. It is, appropriately enough, class. As with everything else in Python, to ensure the code that belongs to the class we are defining is allocated correctly, anything indented under the class line belongs to that class. If we are extending an existing class, we put the parent class or superclass in parentheses after the new class name. This is done below to establish the ApprovalError class as extending the ValueError class, as it is a type of value error, ensuring it can inherit properties and methods from ValueError.

At the start of the python file, immediately after the import statement, enter this:

```
class ApprovalError(ValueError):
```

We need to then set up what methods we are defining, which, like functions, are set up through the def keyword. However, these are indented one level. They also, as mentioned above, "*must*" have the first parameter set aside for the object that each instance of this class belongs to. It is called here, as is typically the case, self. Add these below the class line:

```
→ def __init__(self, errorType, lotID, info1, info2):

→ def createMessage(self):
```

Below these two methods, provide the shell for the Parcel class. I have listed the properties we will have as a comment:

```
class Parcel:


→ def __init__(self, lotID):


#properties--lot ID; lot area; building count; total building area; building type


→ def assess(self):


→ def propose(self, newArea, newType):
```

These set up the blueprints. Now we can use the flow charts to fill in the methods that are defined here.

Task 4: Fill in the Shells of the Classes

We will start with the ApprovalError class. Its constructor, or __init__() method, largely just fills in the properties but uses the other method, createMessage(), to generate the message property. Notice that for the most part, this is done by taking one of the input parameters and simply assigning it to the property, which is distinguished through the self.propertyName syntax. While in this case, I have reused the parameter names as the property names, there is no requirement that they match like this:

```
→ def __init__(self, errorType, lotID, info1, info2):
→ → self.errorType = errorType
→ → self.lotID = str(lotID)
→ → self.info1 = info1
→ → self.info2 = info2
→ → self.message = self.createMessage()
```

The createMessage() method follows the flow chart in a straightforward manner. Mainly, notice that, even though we did not use the self variable in calling the createMessage() method, it is listed as one of the inputs. As mentioned above, it is the first parameter in a method. Otherwise, setting up a method is very similar to setting up a function, including the use of the return keyword to return a value. In this case, it is the message.

Enter it as follows. This will later generate an error, which I have intentionally put in here to illustrate a couple things in Task 5.[5] However, running the code will fail until we finish filling in all of the method shells:

```
→ def createMessage(self):
→ → message = ''
→ → if errorType == 1:
→ → → message = "Your building for " + self.lotID + " is too big to pass the permit process. It currently covers " + self.info1 + "% of the lot and cannot exceed " + self.info2 + "."
→ → elif errorType == 2:
→ → → message = "Your building type for " + self.lotID + " does not match existing buildings on the lot. It is a " + self.info1 + " type while the others are " + self.info2 + "."
→ → else:
→ → → message = "There was an unspecified error in the approval process: type is " + str(self.errorType) + "; lot is " + self.lotID + "; information provided is " + self.info1 + " and " + self.info2 + "."
→ → return message
```

We can now move on to the Parcel class. As mentioned above, the __init__() method will use logic and code largely drawn from the existing calculateFromLot() method. Start by copying the entire contents of the function into the __init__() method. Add an indent to every line so that it is indented where it needs to be.

Now, let's move on to adjusting it. First, the first two lines of the method are no longer needed, as we now have the lot ID established as a required parameter. Instead, though, we need to make sure the object's lotID property is set properly. Do this by replacing the if statement before the query with the following line:

```
self.lotID = lotID
```

I used a slightly different parameter name (lotID versus lotNum). Make the change in the query to refer to the correct variable.

The next lines establishing variables and the building search cursor are largely unchanged. The only changes there is that I have added "FOOT_TYPE" to the end of the list of fields in the variable buildings and added self. to initialize a property called self.buildingCount instead of a generic variable called buildingCount.

---

[5]There's an intentional error in the Parcel class, too, which we will fix.

After these lines, and immediately before the `for building in buildings:` loop, add a line that sets the initial value of the building type property to be `"EMPTY."` If we have any buildings in this lot, it will change. Otherwise, the building type for the lot will remain empty to reflect that there are no buildings in this lot.

Question 3: What is the code for this line, setting the initial value of a buildingType property to "EMPTY"?

The code inside the for loop will only change slightly, as well. We add a line to create a `buildingType` variable which holds the newly added item #3 from the list the cursor gives us inside the loop, and changing the code in the `if` statement to change the building count property of the object, again by using the `self.` syntax, and by changing the new `buildingType` property you initialized as `"EMPTY"` in Question 4 to the building type of this building. The `for` loop should be as follows:

```
→ → for building in buildings:
→ → → buildingBlock = building[0]
→ → → buildingSection = building[1]
→ → → buildingArea = building[2]
→ → → buildingType = building[3]
→ → → if lotBlock == buildingBlock and lotSection == buildingSection:
→ → → → self.buildingCount = self.buildingCount + 1
→ → → → buildingsTotalArea = buildingsTotalArea + buildingArea
→ → → → self.buildingType = buildingType
```

Lastly, we do not have any problems with empty lots anymore, so there is no need to raise a value error when the building count is zero, and there is no need to return any calculations. This is because the purpose of the `__init__()` method is to initialize or construct the variable as it is created. It does not need to return anything. However, there are a couple properties left to set up, so replace these last lines with the following:

```
→ → self.lotArea = lotArea
→ → self.buildingArea = buildingsTotalArea
```

When all is said and done, you should have the following, albeit with the commented line containing the code from Question 4:

```
→ def __init__(self, lotID):
→ → self.lotID = lotID
→ → query = '"KEY" = ' + str(lotID)
→ → search = ARCPY.da.SearchCursor('Lots', ["AREA", "KEY", "BLOCK", "SECTION"], query)
→ → lot = search.next()
→ → lotArea = lot[0]
→ → lotBlock = lot[2]
→ → lotSection = lot[3]
→ → buildings = ARCPY.da.SearchCursor('Buildings', ["BLOCK", "SECTION", "F_AREA", "FOOT_TYPE"])
→ → buildingsTotalArea = 0
→ → self.buildingCount = 0
→ → ###CODE FOR QUESTION 4###
→ → for building in buildings:
→ → → buildingBlock = building[0]
→ → → buildingSection = building[1]
→ → → buildingArea = building[2]
→ → → buildingType = building[3]
→ → → if lotBlock == buildingBlock and lotSection == buildingSection:
→ → → → self.buildingCount = self.buildingCount + 1
→ → → → buildingsTotalArea = buildingsTotalArea + buildingArea
→ → → → self.buildingType = buildingType
→ → self.lotArea = lotArea
→ → self.buildingArea = buildingsTotalArea
```

This is, it so happens, the biggest of the methods. The next method in our code, assess(), is, as you can see in the flow chart, the easiest to implement:

```
→ def assess(self):
→ → buildingRate = self.typeValueDict[self.buildingType]
→ → return self.baseValue * self.lotArea + buildingRate * self.buildingArea
```

You may notice, however, that this method uses a few properties that are not created in the __init__() method. This refers back to the idea above of public and private properties. We can create properties that store variables that are useful for any instance of the class, but are not impacted by the particular parameterization of any instance of this class. Here, the tax rates are not changing based upon the parcel, but every parcel needs to be able to access those rates for the assessment calculations. We can then set them up with a few lines immediately after the class line:

```
class Parcel:


→ typeValueDict = {"COMMERCIAL": 1, "COMMUNITY": 0.3, "EDUCATION": 0.4, "GOVERNMENT": 0.1, "HEALTH":
0.25, "RECREATION/TOURIST": 0.9, "RESIDENTIAL": 0.7}
→ baseValue = 0.1
→ threshold = 30
```

These lines then set up the rates for the buildings based upon their type, as well as the base value rate for the lot size. The last one, threshold, is used within the propose() method to store the maximum percent of the lot which may be occupied by buildings.

The last method we need to fill in here is the propose() method. Enter it as below. Notice that it raises our custom errors to indicate disapproval. It sends up an approval error with type 2 if the building type doesn't match that of an existing building type and the existing type isn't empty. It raises a type 1 error if the percent of the lot covered by buildings will now exceed the predetermined threshold. Note that the information parameters for the ApprovalError are what becomes useful in generating a helpful error message: the new building type and the existing building type for the type 2 error, and the percent and the threshold for the type 1 error. Notice that if there are no approval errors we do not need to reenter the code for calculating the assessment, but instead we just call the self.assess() method:

```
→ def propose(self, newArea, newType):
→ → newTotal = self.buildingArea + newArea
→ → percent = 100 * (newTotal / self.lotArea)
→ → if newType != self.buildingType and self.buildingType != "EMPTY":
→ → → raise ApprovalError(2, self.lotID, newType, self.buildingType)
→ → elif percent > self.threshold:
→ → → raise ApprovalError(1, self.lotID, percent, self.threshold)
→ → else:
→ → → self.buildingType = newType
→ → → self.buildingArea = newTotal
→ → → return self.assess()
```

Now we are ready to try out our code.

Task 5: Debug the Code

First, let's try it out with values which ought to work. Run your tool with the following inputs: lot number 42007001, new building area 100, and new building type RESIDENTIAL.

Open up the message by clicking on the little triangle arrow to the left of the check mark.

Question 4 (10 points): What is the estimated tax assessment of this property?

Now let's look at what happens if the building type doesn't match an existing building. Change the lot number to 42004001, which contains an EDUCATION building already, but leave the new building type as RESIDENTIAL. You should get a massive error message, concluding with NameError: name "errorType" is not defined.

This traceback gives us information on how to find all the different areas where the error might be. We can read it from the bottom up. It tells us that the error was discovered on line 15, in the method called `createMessage`. The line says `if errorType == 1:` and there is nothing called `errorType`. In this situation, that's all you would need to possibly recognize that, while there is a perfectly good property called `errorType`, without `self.`, we are not actually accessing that property. The computer is, instead, looking for a separate variable called `errorType`, which doesn't exist. Therefore, we need to replace `errorType` with `self.errorType`. This is true both here and a couple lines below when we create the message for error type #2.

If we hadn't recognized this in line 15, the error could have been a problem in the method that called `createMessage` instead of in `createMessage` itself. To see what that was, we need to go up one entry in this traceback. It then says it was line 10, within the `__init__` method, that called `createMessage`, and, indeed, you can see that it tells you the line is `self.message = self.createMessage()`.

We can continue up the traceback to see what initialized this error message in the first place. That is in line 65, within the `propose` method, where it raises an `ApprovalError`. While the error is, again, not here, it illustrates one aspect of how custom classes work. We never called anything like `appError = appError.__init__(2, "12180010009", "RESIDENTIAL," "EDUCATION")`, even though, in this line, the `__init__` method is nonetheless called. That is an aspect of creating objects: the `__init__` method is always accessed through what is called the constructor, which is a method whose name is the exact same as the name of the class (`ApprovalError` here) and which has all the input parameters from the __init__() method minus the required `self` parameter. Thus, to construct the object and run the code within the __init__() method, we need to call the constructor through the name of the class, using this piece of code:

```
ApprovalError(2, self.lotID, newType, self.buildingType)
```

to access this method:

```
__init__(self, 2, self⁶.lotID, newType, self.buildingType)
```

Lastly, in the traceback, as this approval error is raised within the `propose()` method, we see that the call to the `propose()` method is on line 80, in the line `newAppraisal = permitParcel.propose(newBuildingSize, newBuildingType)`.

While it is not the case here, it is possible that we could have an error in either the `newBuildingSize` or `newBuildingType` parameters, but it is not discovered until we try to create the error message at the end of this traceback. This will be illustrated later.

Fix the error on lines 15 and 17 by replacing `errorType` with `self.errorType`, if you haven't already done so. Then run the tool again, and you should get an error message, but not the expected one.

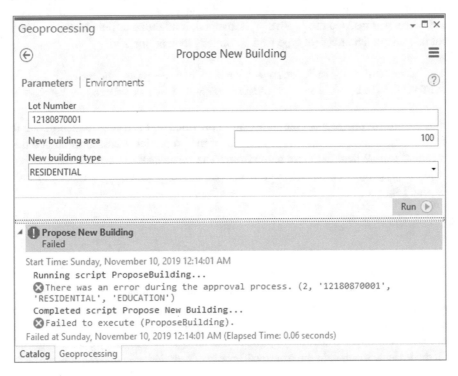

We took the trouble to create a good, custom error message for the user, but it isn't what is shown here. This is because we need to override an existing method, specifically the __str__() method. If you add `ARCPY.AddMessage()` lines, you will see that the `createMessage()` method works perfectly and that the message property is correct. However, we have not told `str(ae)` to actually access that message. Just as `ApprovalError()` calls the __init__() method, str(ae)

---

⁶Part of why the self parameter isn't specified in the ApprovalError() call is shown by how I have set up the __init__() equivalent here. The first self is the self that refers to the ApprovalError object, which is the self that is active in lines 5–21, starting with the ApprovalError class' __init__() method. The other two selfs, in self.lotID and self.buildingType, are the self which is active on line 65, referring to the Parcel object which is raising the ApprovalError. We can conceptually replace self.lotID and self.buildingType with permitParcel.lotID and permitParcel.buildingType within this particular code, although, as I said above in justifying the existence of the self variable in the first place, we have no way of guaranteeing what the actual name of the variable is outside this controlled context.

ᵀʰᵉʳᵉ is, in fact, no explicit variable at this stage which holds the approval error, and none will be created for it until it is caught in the except clause and given to the variable ae. It is stored in a temporary variable until that spot in the code, so the only way we would have to even access it is through the self variable.

calls the __str__() method. Currently, without overriding that method, it instead relies upon a basic __str__() method which just prints the tuple of parameters. We instead want it to print out the error message we have created. This is a very simple addition to the ApprovalError class, defining how we want to implement the __str__() method to override it and return the message we have already created. Define your override of the __str__() method after the __init__() method using the code below:

```
→ def __str__(self):
→ → return self.message
```

Try the tool again with the same inputs, and you should get the correct error message this time.

Now you have an error message saying "There was an error during the approval process. Your building type for 12180870001 does not match existing buildings on the lot. It is a RESIDENTIAL type while the others are EDUCATION."

Lastly, try running the code again, changing the type to EDUCATIONAL and the new building area of 30000, and you should encounter the third error I embedded in the code.[7] You should get the error message below"

```
message = "Your building for " + self.lotID + " is too big to pass the permit process. It currently
covers " + self.info1 + "% of the lot and cannot exceed " + self.info2 + "."
```

Again starting from the bottom, the bug is discovered by the computer as a TypeError in line 18. (Line 20 in my code—depending on how and where you inserted the __str__() method, it might be different for you. The message tells us it is expecting a string, but instead got a number (of type float). So let's look at line 20:

```
message = "Your building for " + self.lotID + " is too big to pass the permit process. It currently
covers " + self.info1 + "% of the lot and cannot exceed " + self.info2 + "."
```

This line is building the error message from the information provided and is trying to incorporate the parameters into a longer string for a helpful error message. This means that one of the properties accessed through the self object is not a string, but this line assumes it is. Where did these come from? The createMessage() method is called within the __init__() method, so the property is established as a string within the __init__() method. Here are the lines for the properties accessed in line 18:

```
self.lotID = str(lotID)
self.info1 = info1
self.info2 = info2
```

From this, we can rule out lotID as the problem, as it is turned into a string here. This means that info1 and/or info2 are not strings. However, they were within the error message for criterion 2, as we got that message without a problem. This means that instead of just turning the info1 and info2 properties into strings here, we might want to handle that earlier in such a manner that info1 and info2 are always presented as strings. This means we want to go and check out the code in the previous line in the traceback, wherein the ApprovalError is raised, which is line 69. It reads as follows:

```
raise ApprovalError(1, self.lotID, percent, self.threshold)
```

The first parameter is the error type, which is not present in line 18, so it cannot be that. We see in the __init__() method that lotID is turned into a string there, so it cannot be that. This leaves just the info1 and info2 properties, which are the percent of the lot covered by buildings, and the maximum permitted percent for any lot. These, it turns out, may well be numbers. Therefore, to fix this, we should turn percent and self.threshold here into strings, replacing them with str(percent) and str(self.threshold).

---

[7] Actually, they are mistakes I made when developing the code for this chapter and I am stepping you through the debugging process I carried out. So this can also be a reminder that, even as you near the end of the semester, you will encounter bugs. I've been programming in various languages for over 20 years (!!!) and still do things like this. The tip for successful programming truly is how to interpret the error messages and not get too frustrated.

Question 5: The tool package will be evaluated by these criteria:
- **Does a tool parameterization which should get approved finish with a regular ArcPy message telling me the estimated appraised value?**
- **Does a tool parameterization with an area too large get denied, with the appropriate error message?**
- **Does a tool parameterization with a type unlike those of existing buildings get denied, with the appropriate error message?**

Task 6: Unguided Work

The town council has passed a new ordinance, restricting the number of building types in each section number. However, a few sections are already being built with more than one type, and they are grandfathered in so that no new types would be permitted. This will entail the development of a new class for the `Neighborhood`.

Please do the following:

6a: Create a new class called `Neighborhood`. The constructor needs to accept a section number as the only input parameter.

6b: Its constructor needs to store the section number to a property.

6c: It's constructor also needs to use a cursor to loop through all the buildings in the section and create a list of property types already present in the section. If there are no buildings, this should be an empty list. (In other words, it should not have `"EMPTY"` in the list.)

6d: Construct an object of the type `Neighborhood` within the `try` block and pass it as a parameter to the parcel's `propose` method.

6e: In the parcel's `propose` method, check to make sure that the new building type does not violate the new ordinance,[8] and raise an appropriate `ApprovalError`, per 7e below, if it does violate it.

6f: Add a new error type, #3, and accept as inputs the error type value 3, the lotID is still the lotID, info1 is the type of the proposed building, and info2 is the list of existing building types, turned into a string.

Example error messages would be:

"Lot 12180950090 cannot accept a COMMERCIAL building because its section already contains RESIDENTIAL buildings."

or

"Lot 12181000002 cannot accept a GOVERNMENT building because its section already contains RESIDENTIAL, HEALTH, and COMMERCIAL buildings."

I will evaluate these using a set of tool parameterizations, some of which are acceptable, and others which should cause this error message to appear.

Save your project as a .ppkx file.

---

[8] If there are no buildings in the section, then any building type is acceptable.

Introductory Comments: Turning Specifications into a Program

This chapter is structured differently from the others but provides an opportunity to develop a simple GIS application all on your own. However, it may also be the most useful if you will use GIS in your job or research. The scenario is one in which you are given specifications and have to create a tool. I provide below an illustrative example of how I would work through that process. While there is no hard and fast way of programming that will be guaranteed to work for everyone, there are a few principles that apply to everyone: the importance of planning and the inevitability of debugging. In the illustrative example, that is, in lieu of the guided tasks, I work through the planning process I would use and show you that, even as someone who first learned programming in 1995, I still make my fair share of bugs.

The general principle is that many of the programming situations you will encounter are unguided, to put it mildly. Imagine being told by a supervisor to write a program to add a tool that does *this* in ArcGIS Pro, whatever *this* may be. They won't tell you what objects and functions you will need. They might not even have the right data for you to evaluate whether the tool works. All you will get is a set of tasks that the tool is supposed to accomplish for the user. In another setting, you might find yourself thinking "it would be better if I had a tool to do *this*," and your observations take the place of the supervisor. Even then, you have no direct guidance, just instructions, and possibly a deadline. What do you do then?

The tools that I use are the ones that we have used throughout the book—pseudocode and flow charts. Others may have different strategies. Regardless of the form it takes, as you work through the tasks at the end of this chapter, I encourage you to take the planning part seriously and move from the plans to the scripts early enough to have time to carry out the inevitable debugging. The following example shows one way of accomplishing this.

Illustrative Example

The files for this example are a GIS file of cities in Europe and a CSV file of airport connections[1] between these cities.

Given these files, suppose I am asked to develop a tool to assist the airport planners in ensuring enough fuel is stocked. They recognize that the amount of fuel needed per day is related to the total length of flights departing the airport. For this task, make the unlikely assumption that there is exactly one flight per day leaving each city for all of its connections. In other words, no routes are flown more than once per day, and they are flown every day. This means we want to sum up the lengths from a specified city to all places connected to that city.

The tool I am to develop will allow the user to type in a city name, read the CSV file, find all the cities connected to the provided city, calculate the sum of distances from the provided city to all of these connected cities, and lastly report this sum of distances within an ARCPY message. Additionally, it should give a sensible error message if that city is not found in the dataset and give a warning if that city has no connections.

With this task, the first thing I would do is take a look at the datasets. The cities file looks straightforward within ArcGIS. It is a point-based file, with a few attributes, although this calculation will need the geometries instead.

---

[1] Not actual connections. I made them up, but think they are pretty plausible, at least as a subset of likely connections.

J. Conley, *A Geographer's Guide to Computing Fundamentals*, Springer Textbooks in Earth Sciences, Geography and Environment, https://doi.org/10.1007/978-3-031-08498-0_17

The connections file initially looks more complicated. Each row is a list of cities. The structure is that the first column is hub city within the airport network, and the remaining columns are the cities connected to that hub. It is set up that most flights are only listed once, because they go from hub to spoke in the network, and to save space, rows for the spokes are not included. For example, the London line has Dundee as a connected location, but Dundee, as a spoke in the network, does not have its own line in the file. Even so, when the user types in Dundee, its connection with London should be included in the calculations.

Having looked at the files to see what I have to work with, I would then start to plan out the tool, laying out the starting and ending points (Fig. 17.1).

Obviously, we need to do more to fill in the black box, but this is a starting point. The next step we might need is to read the connections CSV file. It doesn't depend directly upon getting the city, so I will add this box in without an arrow to it, but it will provide needed information for the (currently) black box calculations (Fig. 17.2).

Once we have the data loaded in, we can start to fill in the black box and, as I have drawn below, make it more gray than black, as the inner workings of the black box get exposed and filled in. The two main components of what has to happen inside this black box are finding the connected cities and adding up all of their distances from the primary city (Fig. 17.3).

We can now break down either of the gray boxes. I'll pick the second one, since it doesn't matter which order we break them down in, as long as they both get done. One thing to think about within your planning is to consider what inputs and outputs you need to get each step done. Adding up all of the distances can be pretty straightforward, although we need to consider what the inputs to this step will be. Naturally, we need the individual distances. But how will the computer represent these distances? There are a few possibilities, and a list or a dictionary comes to mind. I might revise this later, but will, for the time being, choose a list. When adding up the distances, it isn't of great importance which distance corresponds to which connection, and that's the main value of using a dictionary versus a list. Having a list of distances then means looping through all the distances and adding them to a running total. We have done this logic before, so hopefully it is familiar to you (Fig. 17.4).

This makes clear that the other gray box will need to finish by giving us a list of distances. It will end up following a similar logic, but what happens will be more complex. I've started to flesh out what is needed here, making the parts of the black box again clearer, but the lighter gray boxes still have some work to go (Fig. 17.5).

To make the flow chart easier to read, I'll pull out these two gray boxes into their own flow chart. The first part tackles whether the line in the CSV file contains a connection involving the city that the user entered. This should be a simple enough

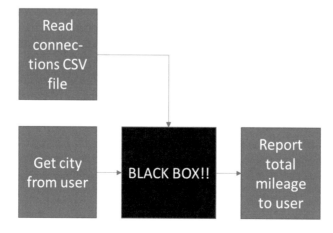

Fig. 17.1 Starting a flow chart

Fig. 17.2 Read a file

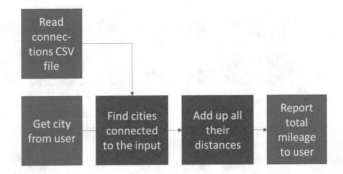

Fig. 17.3 Start working on the black box

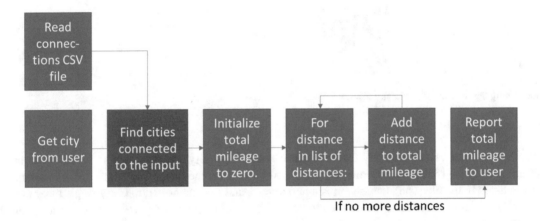

Fig. 17.4 Filling in part of the black box

Fig. 17.5 Adding a loop for getting distances

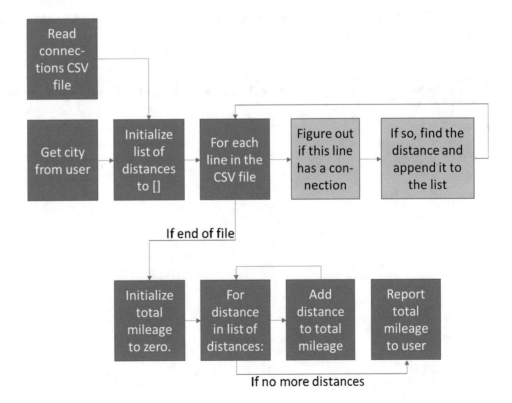

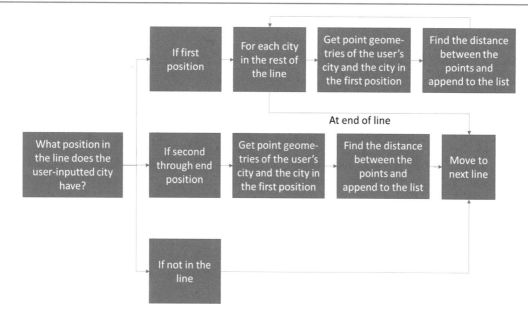

Fig. 17.6 Identifying the connection(s) of a city in a line

check of whether that line includes the city that the user entered. If that city appears, there's at least one connection. If it doesn't appear, there is no connection. Getting the distance(s) correct is the harder part. Here it is important to consider exactly what is listed in the file. The first column is a hub airport, and it is connected to every single other city in the line. However, the city the user entered is not guaranteed to be a hub; it could appear in the first column, or anywhere else in the line, or even nowhere in the line at all. (This last one will be most common.) We want to act differently for these three situations. If the user's city is in the first column, it is connected to every single other city in the line. If it appears in any other column, it is connected to the city in the first column but none of the others. The easiest situation is if it doesn't appear at all. Then we just move on to the next line. Let's put this into a flow chart, then (Fig. 17.6).[2]

We could put this all together into a single flow chart, making sure to embed this logic for a single line into a for loop that cycles through all lines in the CSV file.

---

[2]Hint for later on: this logic, as well as the associated Python code, will be pretty helpful in one of the unguided tasks, where you still want connections, even if not distances.

```
*Untitled*                                                                  —    □    ×

File  Edit  Format  Run  Options  Window  Help

•          connections □ Read in the connections CSV file using the read.csv function
•          userCity □ Input city from the user through the arcpy.GetParameter function.
•          distanceList □ []
•          for line in connections:
○          position □ position of userCity in line
○          if position == 1:
□          for city in line[1:end] :
•          cityGeom □ getGeomOfCity(city)
•          userGeom □ getGeomOfCity(userCity)
•          distance □ cityGeom.distanceTo(userGeom)
#remembering .distanceTo() from the Geometry Exercise.
•          distanceList.append(distance)
○          elif position == -1:
□          This is the city not appearing at all, so do nothing
○          else:
□          city = line[0]
□          cityGeom □ getGeomOfCity(city)
□          userGeom □ getGeomOfCity(userCity)
□          distance □ cityGeom.distanceTo(userGeom)
#remembering .distanceTo() from the Geometry Exercise.
□          distanceList.append(distance)
•          totalMileage □ 0
•          for dist in distanceList:
○          totalMileage □ totalMileage + dist
•          report totalMileage as an Arcpy Message.


•          def getGeomOfCity(cityName):
•          search □ searchCursor(cities file, [SHAPE@, NAME], 'NAME' == cityName)
•          theCity □ search.next()  #assuming exactly one city per name
•          return theCity[0]  #recognizing [0] as the first item in the list for theCity is the geometry.




                                                                        Ln: 8  Col: 14
```

Now we are ready to build our pseudocode. Our flow chart here is detailed enough that the pseudocode will not add a lot more detail, but it will help organize the script when the time comes to write it. The main additional thing I'm doing here is adding in ideas about what functions could be useful. While writing this, I noticed that getting the geometry of a city is going to be a frequent task, so pulled it out as its own function to call whenever I need it:

- connections ← Read in the connections CSV file using the read.csv function
- userCity ← Input city from the user through the arcpy.GetParameter function
- distanceList ← []
- for line in connections:[3]
  - position ← position of userCity in line
  - if position == 1:
    - for city in line[1:end][4]:
      - cityGeom ← getGeomOfCity(city)
      - userGeom ← getGeomOfCity(userCity)
      - distance ← cityGeom.distanceTo(userGeom)
        #remembering .distanceTo() from the Geometry Exercise.
      - distanceList.append(distance)

---

[3] I had considered creating a custom object or a dictionary to hold both the city name and its connections as a list. This turned out to be overkill here but might in fact be useful in one or more of the unguided tasks.

[4] Recall from Chap. 2 the syntax of a subset of a list. That's what I'm working with here.

- elif position == −1:
  - This is the city not appearing at all, so do nothing
- else:
  - city = line[0]
  - cityGeom ← getGeomOfCity(city)
  - userGeom ← getGeomOfCity(userCity)
  - distance ← cityGeom.distanceTo(userGeom)
- #remembering .distanceTo() from the Geometry Exercise.
  - distanceList.append(distance)
- totalMileage ← 0
- for dist in distanceList:
  - totalMileage ← totalMileage + dist
- report totalMileage as an Arcpy Message.

- def getGeomOfCity(cityName):
  - search ← searchCursor(cities file, [SHAPE@, NAME], 'NAME' == cityName)
  - theCity ← search.next() #assuming exactly one city per name
  - return theCity[0] #recognizing [0] as the first item in the list for theCity is the geometry.

Having finished this, it is now time to build the code.

You might have observed throughout this course that you never actually started a Python file of your own. This is strictly because accessing the appropriate IDLE editor is not nearly as straightforward as it should be. If you look in the Windows start menu, under ArcGIS, you will indeed see an option labeled "IDLE (Python GUI)." This is incorrect on two accounts: it is the command line interface, or the shell, not the file editor; and it is the version of Python with ArcGIS Desktop (Python 2.x) instead of the version for ArcGIS Pro (Python 3.x).

So instead of this, what I do is open up an existing file by right-clicking on it and choosing "Edit with IDLE (ArcGIS Pro)." Then, once the correct version of IDLE is up and running, go to the File menu and click New File. Now you can close the one you started from, and get your new file going.

To me, the first thing is to copy the pseudocode over into the new file and make it all comments. This way, we can then fill in beneath each comment with Python code to carry out that comment. Of course, it starts out looking pretty bad.

```
# connections <-- Read in the connections CSV file using the read.csv function
# userCity <-- Input city from the user through the arcpy.GetParameter function.
# distanceList <-- []
# for line in connections:
# --> position <-- position of userCity in line
# --> if position == 1:
# --> --> for city in line[1:end] :
# --> --> --> cityGeom <-- getGeomOfCity(city)
# --> --> --> userGeom <-- getGeomOfCity(userCity)
# --> --> --> distance <-- cityGeom.distanceTo(userGeom) #remembering .distanceTo() from the Geometry Exercise
# --> --> --> distanceList.append(distance)
# --> elif position == -1:
# --> --> This is the city not appearing at all, so do nothing
# --> else:
# --> --> city = line[0]
# --> --> cityGeom <-- getGeomOfCity(city)
# --> --> userGeom <-- getGeomOfCity(userCity)
# --> --> distance <-- cityGeom.distanceTo(userGeom) #remembering .distanceTo() from the Geometry Exercise.
# --> --> distanceList.append(distance)
# --> totalMileage <-- 0
# --> for dist in distanceList:
# --> --> totalMileage <-- totalMileage + dist
# --> report totalMileage as an Arcpy Message.


# def getGeomOfCity(cityName):
# --> search <-- searchCursor(cities file, [SHAPE@, NAME], 'NAME' == cityName)
# --> theCity <-- search.next()  #assuming exactly one city per name
# --> return theCity[0]  #recognizing [0] as the first item in the list for theCity is the geometry.
```

But we can change all these unusual symbols from the MS Word outline into better-looking comments.

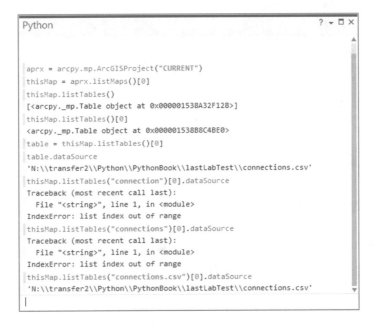

I fill in the code now, referring to some of the previous exercises, especially those which help me remember how reading a CSV file works, how search cursors work, and how geometries work.

After that, this is what I have:

```
import csv
import arcpy

# connections <-- read in the connections CSV file using the read.csv function


connections_path = "D:\books\PythonManual\Chapter17\connections.csv"
connections_file = open(connections_path)
connections_reader = csv.reader(connections_file)

#userCity <-- Input city from the user through the arcpy.GetParameter function
userCity = arcpy.GetParameter(0)

# distanceList <-- []
distanceList = []

# for line in connections
for line in connections_reader:

# --> position <-- position of userCity in line
    position = line.index(userCity)

# --> if position == 1: NOTE: I CORRECTED THE 1 TO 0 IN THE CODE
    if position == 0:

# --> --> for city in line[1:end] :
        for city in line[1:]

# --> --> --> cityGeom <-- getGeomOfCity(city)
            cityGeom = getGeomOfCity(city)

# --> --> --> userGeom <-- getGeomOfCity(userCity)
            userGeom = getGeomOfCity(userCity)

# --> --> --> distance <-- cityGeom.distanceTo(userGeom) # remembering .distanceTo() from the
Geometry Exercise
            distance = cityGeom.distanceTo(userGeom)

# --> --> --> distanceList.append(distance)
            distanceList.append(distance)

# --> elif position == -1:
    elif position == -1:

# --> --> This is the city not appearing at all, so do nothing


# --> else:
    else:

# --> --> city = line[0]
```

```
        city = line[0]

# --> --> cityGeom <-- getGeomOfCity(city)
        cityGeom = getGeomOfCity(city)

# --> --> userGeom <-- getGeomOfCity(userCity)
        userGeom = getGeomOfCity(userCity)

# --> --> distance <-- cityGeom.distanceTo(userGeom) # remembering .distanceTo() from the Geometry
Exercise
        distance = cityGeom.distanceTo(userGeom)

# --> --> distanceList.append(distance)
        distanceList.append(distance)

# --> totalMileage <-- 0
    totalMileage = 0

# --> for dist in distanceList:
    for dist in distanceList:

# --> --> totalMileage <-- totalMileage + dist
        totalMileage = totalMileage + dist
# --> report totalMileage as an Arcpy Message
    arcpy.AddMessage("The total mileage is " + totalMileage)

# def getGeomOfCity(cityName):
# --> search <-- searchCursor(cities file, [SHAPE@, NAME], 'NAME' == cityName)
# --> theCity <-- search.next() #assuming exactly one city per name
# --> return theCity[0] #recognizing [0] as the first item in the list for theCity is the
geometry.

def getGeomOfCity(cityName):
    query = '"CITY_NAME" = \'' + cityName + "'"
    search = arcpy.da.searchCursor('cities', ['SHAPE@', 'CITY_NAME'], query)
    theCity = search.next()
    return theCity[0]
```

Now that this is set up (although I'll certainly need to debug it), it's time to set up the tool. I go to ArcGIS Pro, create a new tool in my toolbox, add a string parameter called City Name, and connect it to this script.

To try it out, I type in London and click the Run button. Not surprisingly, it failed.

If I view the details, it tells me that I have a syntax error. I forgot the colon at the end of the line:

```
File "D:\books\PythonManual\Chapter17\Chap17.py", line 26
    for city in line[1:]
                        ^
SyntaxError: invalid syntax
```

I correct that and run it again, and get a new error:

```
File "D:\books\PythonManual\Chapter17\Chap17.py", line 46
    else:
        ^
IndentationError: expected an indented block
```

My approach to ensuring the code does nothing when the city is not in the list of having an `elif` clause that contains only comments did not work. It expects to do something.[5] A quick Google search alerts me to the `continue` keyword, which might work here. I put this keyword as the only statement inside the `elif` clause. Saving and running at least gives a different, and later, bug:

```
File "D:\books\PythonManual\Chapter17\Chap17.py", line 62
  distanceList.append(distance)
  ^
IndentationError: unexpected indent
```

I go to line 62 and indeed find that the line there had one extra space in it, throwing off the indentation. I remove that space, save, and run again:

```
Traceback (most recent call last):
  File "D:\books\PythonManual\Chapter17\Chap17.py", line 7, in <module>
  connections_file = open(connections_path)
OSError: [Errno 22] Invalid argument: 'D:\x08ooks\\PythonManual\\Chapter17\\connections.csv'
```

Having moved from an error on line 62 to an error on line 7 generally indicates that the syntax errors are all completed, and having started over at the beginning, it attempted to run the code and encountered this problem. Therefore, I am moving from syntax errors to runtime errors, which is definitely progress! It didn't like my file path for the connections file, because I had copied and pasted from Windows Explorer without either doubling up the slashes or changing them to backslashes. I fix that and run again:

```
Traceback (most recent call last):
  File "D:\books\PythonManual\Chapter17\Chap17.py", line 29, in <module>
  cityGeom = getGeomOfCity(city)
NameError: name 'getGeomOfCity' is not defined
```

Here, it is complaining that the computer doesn't know what `getGeomOfCity` is, although we do define it as a function. The problem is that it tries to run the code one line at a time, and it is defined after I try to use it, so I'll need to move its definition further up. I move the entire function from the end to right after the import statements, save, and run again:

```
Traceback (most recent call last):
  File "D:\books\PythonManual\Chapter17\Chap17.py", line 41, in <module>
    cityGeom = getGeomOfCity(city)
  File "D:\books\PythonManual\Chapter17\Chap17.py", line 11, in getGeomOfCity
    search = arcpy.da.searchCursor('cities', ['SHAPE@', 'CITY_NAME'], query)
AttributeError: module 'arcpy.da' has no attribute 'searchCursor'
```

This time, I find another typo. The class SearchCursor has a capital S, not a lowercase s, at the start. I fix that and try again:

```
Traceback (most recent call last):
  File "D:\books\PythonManual\Chapter17\Chap17.py", line 41, in <module>
    cityGeom = getGeomOfCity(city)
  File "D:\books\PythonManual\Chapter17\Chap17.py", line 12, in getGeomOfCity
    theCity = search.next()
StopIteration
```

---

[5]This is not true of all programming languages and is therefore an example of how trying a technique that works in one language can occasionally fail in another.

This one took some more investigation. In line 41, it is calling the connection city from the file, not the one the user entered, so a typo in the name the user entered cannot do this. By giving a `StopIteration` error (which we might want to go in later and catch, so that we can have a better error message), Python suggests that the search cursor returned nothing. This could mean the query is wrong or that the connections file has a typo in it. To investigate, I enter an `arcpy.AddMessage()` statement to see what the query is:

```
def getGeomOfCity(cityName):
 query = '"CITY_NAME" = \'' + cityName + "'"
 arcpy.AddMessage("query is " + query)
 search = arcpy.da.SearchCursor('cities', ['SHAPE@', 'CITY_NAME'], query)
 theCity = search.next()
 return theCity[0]
```

It dutifully prints out each of the queries until it looks for Copenhagen, which suggests that the query that returned nothing was Copenhagen:

```
Start Time: Friday, October 8, 2021 3:47:39 PM
query is "CITY_NAME" = 'Dublin'
query is "CITY_NAME" = 'London'
query is "CITY_NAME" = 'Birmingham'
query is "CITY_NAME" = 'London'
query is "CITY_NAME" = 'Manchester'
query is "CITY_NAME" = 'London'
query is "CITY_NAME" = 'Edinburgh'
query is "CITY_NAME" = 'London'
query is "CITY_NAME" = 'Glasgow'
query is "CITY_NAME" = 'London'
query is "CITY_NAME" = 'Reykjavik'
query is "CITY_NAME" = 'London'
query is "CITY_NAME" = 'Paris'
query is "CITY_NAME" = 'London'
query is "CITY_NAME" = 'Frankfurt am Main'
query is "CITY_NAME" = 'London'
query is "CITY_NAME" = 'Berlin'
query is "CITY_NAME" = 'London'
query is "CITY_NAME" = 'Oslo'
query is "CITY_NAME" = 'London'
query is "CITY_NAME" = 'Copenhagen'
Failed script Calculate Connection Distances...
```

Further investigation points out that there is an error in the connections file. If you look at the cities file, you'll notice that local spellings are used instead of consistent English spellings. Therefore, the capital of Denmark is supposed to use the Danish spelling "Kobenhavn," not the English Spelling "Copenhagen." I fix this and try again. That fixed the issue above, so I commented out the message and ran it again to better see the next error:

```
Traceback (most recent call last):
 File "D:\books\PythonManual\Chapter17\Chap17.py", line 86, in <module>
 arcpy.AddMessage("The total mileage is " + totalMileage)
TypeError: can only concatenate str (not "float") to str
```

Ah. I forgot to turn the total mileage into a string for the message. You might notice by now that most of these errors are pretty simple, and I can assure you that they were not intentionally placed when I drafted the script. Even an experienced programmer can make simple bugs like this. I fix this and move on:

```
The total mileage is 511.6431334882917
The total mileage is 517.998799729663
The total mileage is 521.6877135451559
The total mileage is 526.8371868082567
The total mileage is 535.7869672414591
The total mileage is 549.3316331879395
The total mileage is 563.1155387492284
The total mileage is 576.5286862872425
The total mileage is 596.369984461331
The total mileage is 627.9438187180415
The total mileage is 666.060851663013
The total mileage is 696.7585748499515
The total mileage is 717.9619706191115
The total mileage is 732.6620673900278
The total mileage is 763.6511295088417
The total mileage is 790.9762742053769
The total mileage is 810.6535182935246
The total mileage is 821.786287185185
The total mileage is 837.7095753378401
The total mileage is 849.3004462805711
The total mileage is 859.6964371739018
The total mileage is 875.2817716296667
The total mileage is 926.5070868779868
The total mileage is 971.255393840486
Failed script Calculate Connection Distances...
Traceback (most recent call last):
 File "D:\books\PythonManual\Chapter17\Chap17.py", line 33, in <module>
 position = line.index(userCity)
ValueError: 'London' is not in list
Failed to execute (CalculateConnectionDistances).
```

Ok. This is odd, and perhaps moving into logical error territory. It reported the total mileage many times, which is itself a bit disconcerting. And maybe the numbers are strangely low, but we will get to that later. The ValueError is the bigger problem here. It seems to have a problem with returning the index of something that is not in the list. So I'll have to readjust some of the logic there, using the in keyword, rather than the seeing if the position is -1. I cannot just change the elif condition to elif city in line == False or something like that because the conditional statement to notice that it isn't in the list must occur before line 33. Otherwise, when the computer gets to line 33, it is still a problem. In fact, with this, the elif portion of the conditional statement is not needed at all. There are several ways to address this, and what follows is what I ended up with, after also fixing indentations to reflect the new if statement:

```
# --> position <-- position of userCity in line
    if userCity in line:
        position = line.index(userCity)


# --> if position == 1: NOTE: I CORRECTED THE 1 TO 0 IN THE CODE
        if position == 0:


# --> --> for city in line[1:end] :
            for city in line[1:]:


# --> --> --> cityGeom <-- getGeomOfCity(city)
                cityGeom = getGeomOfCity(city)


# --> --> --> userGeom <-- getGeomOfCity(userCity)
                userGeom = getGeomOfCity(userCity)
```

```
# --> --> --> distance <-- cityGeom.distanceTo(userGeom)
# remembering .distanceTo() from the Geometry Exercise
                distance = cityGeom.distanceTo(userGeom)


# --> --> --> distanceList.append(distance)
                distanceList.append(distance)


# --> else:
        else:


# --> --> city = line[0]
            city = line[0]


# --> --> cityGeom <-- getGeomOfCity(city)
            cityGeom = getGeomOfCity(city)


# --> --> userGeom <-- getGeomOfCity(userCity)
            userGeom = getGeomOfCity(userCity)


# --> --> distance <-- cityGeom.distanceTo(userGeom)
# remembering .distanceTo() from the Geometry Exercise
            distance = cityGeom.distanceTo(userGeom)


# --> --> distanceList.append(distance)
            distanceList.append(distance)
```

This time, it passed. Hooray! But the message reports the total mileage many times, which is unusual and is low if we are trying to estimate total miles or kilometers into and from what is, at this entry, a busy hub of London. First, the multiple reporting problem is easy to resolve. Looking closely, I accidentally indented the `arcpy.AddMessage()` function too far, so it reports for every single line of the connections file. I'll unindent that a level and rerun the code:

```
The total mileage is 971.255393840486
```

Much better, but still with a suspiciously low number. The `distanceTo()` method uses the coordinate system of the input data, which, in this case, is decimal degrees. We want, instead, to use something that gives a unit that is consistent globally, like meters or miles. I look into the different geometry methods and find `angleAndDistanceTo()` which, according to the documentation, might be useful, because it returns "distance (in meters)" as part of its return value. I'll try it, changing the distance portion to what you see below in both times the line was used:

```
distance = cityGeom.angleAndDistanceTo(userGeom)[1] / 1000
```

Running the tool then gave this output:

```
The total mileage is 75670.04804482049
```

This is a much better result, although it looks like we might want to fix the output message, as the word "mileage" might not work for kilometers! I also put in a `round` function to limit the digits that are presented:

```
arcpy.AddMessage("The total distance is " + str(round(totalMileage)) + " km.")
```

Running this again gives the following:

```
The total distance is 75670 km.
```

It isn't obvious, but there is another logical error embedded in here. Flights between hubs are listed twice. That means that, say, the flight from London to Paris is counted twice, once when it reads the London line and again when it reads the Paris line. We can, in fact, fix this with a little bit of creativity and a keyword that I haven't formally introduced but which some of you may have found anyway: `break`. This keyword forces the code to leave a Python loop no matter what. What I'll do here is to leave the code alone if a city isn't the hub, because the error only arises for connections between two hub cities. I will, though, have to change how it works for hubs. Since all of the hub's flights are listed in that line, the only information I need for a hub is right there, in that line. So if and when I encounter a hub, I'll reset the distance list to empty, append everything within that line of the CSV file and then break the loop to ensure I don't add anything more to it. This ultimately ensures that only the distances within this line contribute to the total for a hub:

```
# --> if position == 1: NOTE: I CORRECTED THE 1 TO 0 IN THE CODE
        if position == 0:
            distanceList = []
# --> --> for city in line[1:end] :
            for city in line[1:]:

# --> --> --> cityGeom <-- getGeomOfCity(city)
                cityGeom = getGeomOfCity(city)

# --> --> --> userGeom <-- getGeomOfCity(userCity)
                userGeom = getGeomOfCity(userCity)

# --> --> --> distance <-- cityGeom.distanceTo(userGeom)
# remembering .distanceTo() from the Geometry Exercise
                distance = cityGeom.angleAndDistanceTo(userGeom)[1] / 1000

# --> --> --> distanceList.append(distance)
                distanceList.append(distance)
            break
```

Fixing this revealed another error, in which the variable `totalMileage` is not defined when it is needed. This seems unusual:

```
Traceback (most recent call last):
 File "D:\books\PythonManual\Chapter17\Chap17.py", line 81, in <module>
 arcpy.AddMessage("The total distance is " + str(round(totalMileage)) + " km.")
NameError: name 'totalMileage' is not defined
```

Going to this line in the code, we see there is an indentation problem. It is indented one level, so is therefore in the code for the `for` loop. Breaking out of the `for` loop when we did, because London is the hub for the first line of the CSV file, means that the remainder of the loop never got run, even once. This means that `totalMileage` was never initialized and the `for` loop after its initialization was never done. This means I need to unindent `totalMileage` initialization the second `for` loop to ensure they are not inside the first `for` loop. After all, I want to run the total distance calculation after appending everything and looking at the entire CSV file, not inside the loop running the CSV file:

```
The total distance is 40230 km.
```

Even though this is better, we aren't done testing just yet. Software should be tested by a series of checks, so just because it passed on one parameterization, doesn't guarantee it will pass with all possible parameterizations. It would still be good to see what happens for a city that is in the connections file, but not as a hub, as well as testing what happens for a city that is in the shapefile, but not the connections file, and a city that just doesn't exist.

I'll now try Vienna as a city that is in the connections file, but not as a hub:

```
The total distance is 2156 km.
```

It seems to run correctly, giving a smaller amount. Now for Malmo, which is a Swedish city near Kobenhavn, but is not listed in this set of connections:

```
The total distance is 0 km.
```

It gives zero, which is good. Lastly, I'll try Lodnon, which, as a misspelled version of London, isn't in the cities file:

```
The total distance is 0 km.
```

It also reported zero, which is probably not what we wanted. The only time this would raise an error is when running the search cursor in getGeomOfCity, but we only run that function when we see a city in the connections file. If I move that to getting the userGeom variable as soon as I read the parameter, I could catch this problem. It also means putting that into a try/except block. Recall that searching for something that doesn't exist gives a StopIteration exception, so we will use this information to catch exactly what we need. We can have the try/except statement be compact; it doesn't have to try everything when we know the exception we are interested in would be generated in a specific line:

```
#userCity <-- Input city from the user through the arcpy.GetParameter function
userCity = arcpy.GetParameter(0)

# userGeom <-- getGeomOfCity(userCity)
try:
 userGeom = getGeomOfCity(userCity)
except StopIteration as si:
  arcpy.AddError("Your city is not in the database. Please check your spelling and/or continent.")
```

Running it again with Lodnon gave the appropriate error message:

```
Your city is not in the database. Please check your spelling and/or continent.
```

However, if you look at the messages, it still reports the total distance as 0 km. This is because the AddError() function does not stop the entire program but simply gives an error message and then moves on with the rest of the program. We can easily enough remedy this, by setting a Boolean variable while in the except clause, and only reporting the final message if that indicates there was no error:

```
# userGeom <-- getGeomOfCity(userCity)
noError = True
try:
 userGeom = getGeomOfCity(userCity)
except StopIteration as si:
  arcpy.AddError("Your city is not in the database. Please check your spelling and/or continent.")
 noError = False

...


# --> report totalMileage as an Arcpy Message
if noError:
 arcpy.AddMessage("The total distance is " + str(round(totalMileage)) + " km.")
```

I save this and get a better result, giving the error message only, without the extra zero distance claim.

The last update I'll make is to have the code be able to find the connections table as long as it is loaded into the ArcGIS Project, rather than have the file path hard-coded into the script. First, I try out a few lines in the ArcGIS Python window. This allows rapid debugging, and at the end, I have the line I need to be able to have the file path for the connections file:

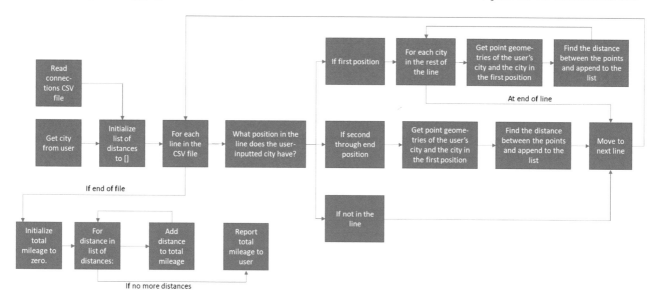

```
# connections <-- read in the connections CSV file using the read.csv function
```

```
aprx = arcpy.mp.ArcGISProject("CURRENT")
thisMap = aprx.listMaps()[0]
connections_path = thisMap.listTables("connections.csv")[0].dataSource
connections_file = open(connections_path)
connections_reader = csv.reader(connections_file)
```

I try it with Madrid. It gives an error, indicating another problem. It is the same `StopIteration` problem that we had above when the search query returned nothing, so I uncomment out the `arcpy.AddMessage()` line telling me the query. Look at the end, and notice it is trying to find a city whose name is an empty string. Not all rows are the same length, and at least loading it into ArcGIS, it fills in the shorter rows with empty strings. There are, of course, no cities with no name:

```
Start Time: Friday, October 8, 2021 4:51:10 PM
query is "CITY_NAME" = 'Madrid'
query is "CITY_NAME" = 'London'
query is "CITY_NAME" = 'Paris'
query is "CITY_NAME" = 'Amsterdam'
query is "CITY_NAME" = 'Frankfurt am Main'
query is "CITY_NAME" = 'Berlin'
query is "CITY_NAME" = 'Stockholm'
query is "CITY_NAME" = 'Praha'
query is "CITY_NAME" = 'Athinai'
query is "CITY_NAME" = 'Budapest'
query is "CITY_NAME" = 'Milano'
query is "CITY_NAME" = 'Roma'
query is "CITY_NAME" = 'London'
query is "CITY_NAME" = 'Paris'
query is "CITY_NAME" = 'Amsterdam'
```

```
query is "CITY_NAME" = 'Frankfurt am Main'
query is "CITY_NAME" = 'Berlin'
query is "CITY_NAME" = 'Stockholm'
query is "CITY_NAME" = 'Praha'
query is "CITY_NAME" = 'Athinai'
query is "CITY_NAME" = 'Budapest'
query is "CITY_NAME" = 'Milano'
query is "CITY_NAME" = 'Roma'
query is "CITY_NAME" = 'Barcelona'
query is "CITY_NAME" = 'Lisboa'
query is "CITY_NAME" = 'Santander'
query is "CITY_NAME" = 'Pamplona'
query is "CITY_NAME" = 'Sevilla'
query is "CITY_NAME" = 'Valencia'
query is "CITY_NAME" = ''
Failed script Calculate Connection Distances...
Traceback (most recent call last):
 File "D:\books\PythonManual\Chapter17\Chap17.py", line 53, in <module>
    cityGeom = getGeomOfCity(city)
 File "D:\books\PythonManual\Chapter17\Chap17.py", line 13, in getGeomOfCity
    theCity = search.next()
StopIteration
Failed to execute (CalculateConnectionDistances).
```

This means we need to add another check to avoid generating this error. Here is the for loop for going through the cities connected to hubs, with the if statement included and the subsequent lines indented appropriately:

```
# --> --> for city in line[1:end] :
          for city in line[1:]:
              if city != '':
# --> --> --> cityGeom <-- getGeomOfCity(city)
                  cityGeom = getGeomOfCity(city)

# --> --> --> distance <-- cityGeom.distanceTo(userGeom)
# remembering .distanceTo() from the Geometry Exercise
                  distance = cityGeom.angleAndDistanceTo(userGeom)[1] / 1000

# --> --> --> distanceList.append(distance)
                  distanceList.append(distance)
```

This now allows Madrid to run without any errors:

```
The total distance is 20719 km.
```

Starting below, you can see the entire code that I ended with. As you can see, this was not a simple process. This is why, as I mentioned earlier, you will want to start by planning things, and not put it off until the last minute. At the same time, don't be afraid to deviate slightly from the plan, as I did a couple times, to fix bugs. Likewise, having a plan for checking the tool under a series of inputs that the user might make, whether they are correct or incorrect. In this example, I did not stop testing the tool when it worked for London and, indeed, continuing to test a few other cities found more bugs. You should be equally willing to test the tools under a range of scenarios to better ensure they will work:

```
import csv
import arcpy
```

```
# def getGeomOfCity(cityName):
# --> search <-- searchCursor(cities file, [SHAPE@, NAME], 'NAME' == cityName)
# --> theCity <-- search.next() #assuming exactly one city per name
# --> return theCity[0] #recognizing [0] as the first item in the list for theCity is the
geometry.

def getGeomOfCity(cityName):
 query = '"CITY_NAME" = \'' + cityName + "'"
 arcpy.AddMessage("query is " + query)
 search = arcpy.da.SearchCursor('cities', ['SHAPE@', 'CITY_NAME'], query)
 theCity = search.next()
 return theCity[0]


# connections <-- read in the connections CSV file using the read.csv function


aprx = arcpy.mp.ArcGISProject("CURRENT")
thisMap = aprx.listMaps()[0]
connections_path = thisMap.listTables("connections.csv")[0].dataSource
connections_file = open(connections_path)
connections_reader = csv.reader(connections_file)

#userCity <-- Input city from the user through the arcpy.GetParameter function
userCity = arcpy.GetParameter(0)

# userGeom <-- getGeomOfCity(userCity)
noError = True
try:
 userGeom = getGeomOfCity(userCity)
except StopIteration as si:
  #arcpy.AddError("Your city is not in the database. Please check your spelling and/or
continent.")
 noError = False

# distanceList <-- []
distanceList = []

# for line in connections
for line in connections_reader:

# --> position <-- position of userCity in line
 if userCity in line:
 position = line.index(userCity)

# --> if position == 1: NOTE: I CORRECTED THE 1 TO 0 IN THE CODE
        if position == 0:
            distanceList = []
# --> --> for city in line[1:end] :
            for city in line[1:]:
                if city != '':
# --> --> --> cityGeom <-- getGeomOfCity(city)
                    cityGeom = getGeomOfCity(city)
```

```
# --> --> --> distance <-- cityGeom.distanceTo(userGeom)
# remembering .distanceTo() from the Geometry Exercise
                    distance = cityGeom.angleAndDistanceTo(userGeom)[1] / 1000


# --> --> --> distanceList.append(distance)
                    distanceList.append(distance)
          break
# --> else:
      else:


# --> --> city = line[0]
          city = line[0]


# --> --> cityGeom <-- getGeomOfCity(city)
          cityGeom = getGeomOfCity(city)


# --> --> distance <-- cityGeom.distanceTo(userGeom)
# remembering .distanceTo() from the Geometry Exercise
          distance = cityGeom.angleAndDistanceTo(userGeom)[1] / 1000


# --> --> distanceList.append(distance)
          distanceList.append(distance)


# --> totalMileage <-- 0
totalMileage = 0

# --> for dist in distanceList:
for dist in distanceList:

# --> --> totalMileage <-- totalMileage + dist
    totalMileage = totalMileage + dist
# --> report totalMileage as an Arcpy Message
if noError:
    arcpy.AddMessage("The total distance is " + str(round(totalMileage)) + " km.")
```

Unguided Work

I will expect to see four tools in the toolbox for this project, with the specifications in the next section.

Evaluation Criteria:
- Was the program for the tool planned beforehand, using a flow chart, pseudocode, or something else?
- Each new tool and its dialog box are assembled appropriately.
- The first task in each tool works properly.
- The second task in each tool works properly.
- There are comments explaining what you did.

Specifications for Tools

(1) City Selection

For the first tool, you will create an application that allows the user to label a particular feature. In this case, you will work with a shapefile of cities in Europe. Your application should allow the user to label the specific city that s/he has selected.

Your application should include at least the following functionality:

- Get a city name from the user, and use a custom label class with a query that isolates only that city to label only that city with its name.
- Create a symbology for the cities layer with the following specifications:
  - It is a graduated symbol, using the POP_RANK field, with seven classes, equal interval, with symbol sizes from 4 (minimum) to 15 (maximum).
  - The symbol template is a blue circle.

I want the user to be able to enter a city into the dialog box and have the map zoom to that city. If the user enters a city name that isn't in the database (e.g., misspelling London as Lodnon), it should warn the user nicely rather than crash ArcGIS or give a hard-to-read ArcGIS error message.

(2) Median Calculation

ArcGIS provides a nice "summary statistics" tool, although it does not have everything we might want. In particular, it is lacking the median, which would be the most appropriate statistic for the ordinal "POP_RANK" variable. Your second tool should allow users to select a country and tell them (via an ArcPy Message) the median POP_RANK value for that country.

Your application should include at least the following functionality:

- Get a country name from the user and use a query to access just the cities inside that country.
- Calculate the median POP_RANK of all cities in that country and report the median to the user.

Calculating the median will involve forming a list of all POP_RANK values for that country, sorting it, and finding the middle value in that list.

(3) Print Connections

During disease outbreaks, infectious diseases can often spread through the airline network. Using the .csv file of connections, allow the user to input a city where the outbreak started, and present messages with two lists of cities. The first contains the cities which are of primary concern, because they are immediate connections with the city that the user selects.

The second list contains cities of secondary concern, because they are connected with a city of primary concern, but not the initially selected location.

For example, if the outbreak started in Vienna, the cities of primary concern are Frankfurt, Berlin, Warsaw, Praha, and Budapest, because they are directly connected to Vienna.

Every place in the five rows for Frankfurt, Berlin, Warsaw, Praha, and Budapest then becomes a city of secondary concern, because the disease could have gone from Vienna through Berlin to Dresden, for example, making Dresden a city of secondary concern. There are 93 connections involved in secondary cities, although there are fewer than 93 cities of secondary concern, because many of the connections overlap. For example, Both Berlin and Frankfurt connect to Oslo, so Oslo is a city of secondary concern, but it only needs to appear once in the list.

Your application should include at least the following functionality:

- Get a city name from the user and use the .csv file to identify primary and secondary cities of concern as defined above.
- Report, through Arcpy messages, both lists. If the entered city has no connections, there should be a message stating this. For the list of cities of secondary concern, no cities of primary concern should appear, nor should the initially selected city. Also, no city should be listed twice, even if it has two connections with separate cities of primary concern.

If the user enters a city name that isn't in the database (e.g., misspelling London as Lodnon), it should warn the user nicely rather than crash ArcGIS or give a hard-to-read ArcGIS error message.

(4) Suggest New Connections

An airline might want to see if they can expand their network. Your intent in this task is to allow the user to enter a city name and identify any cities of population rank 6 or lower (as lower ranks indicate larger cities) with distances between 100 km and 500 km of the selected city.

Your application should include at least the following functionality:

- Get a city name from the user, identify the cities within the 100 to 500 km range, and use the .csv file to identify those which lack connections.
- Report, through Arcpy messages, a list of cities which need connections, presented in order of the rank with the largest cities presented first.

If the user enters a city name that isn't in the database (e.g., misspelling London as Lodnon), it should warn the user nicely rather than crash ArcGIS or give a hard-to-read ArcGIS error message.

Part IV

Sharing Your Work with Others

Even though it can often seem to be a very solitary practice, computer programming is greatly enhanced when connected with the community of millions of programmers worldwide. You have already, perhaps unwittingly, worked within some of these connections. It can even be as simple as searching online to investigate whether anyone else has encountered the bug you are currently staring at, and seeing if their solution will work for you, too. Also, any time you import a package, or even use software written by others, you are building upon the work of others. This sharing is made more explicit with WebGIS, another subject which readily fills its own book, and is thus not fully explored here. The primary ways that sharing happens is through connections with packages others have written, and creating your own package to share with others.

Now that you've built a tool, this section expands upon these skills to enable the sharing of content with the wider world. It starts off with sharing maps through the use of ArcPy for Web GIS using ArcGIS Online and the `arcpy.sharing` module. Then, the connection with other packages is provided, with the use of `Numpy` procedures, as well as being able to create tools that run off of R code instead of Python. The last two chapters focus on the development of your own package. Since most packages contain more than one Python file, Chap. 20 is about the creation and management of programs with more than one file. Chapter 21 builds upon this to create your own package and connect it with ArcGIS Pro.

ArcPy for Web GIS Using ArcGIS Online

This will introduce concepts of Web GIS and illustrate these concepts through the use the `arcpy.sharing` module. The use of data portals for Web GIS is also included. Just as important as writing the code here is creating metadata and the necessary information to enable people who come across your ArcGIS Online project to use and interpret it effectively.

Introductory Comments: Shared Services

The first stage of sharing projects with the broader world often comes through the use of preexisting shared services. These services can be used to exchange data, both sharing data that you have collected and accessing data from other sources without having to download it directly. These services can also be used to share more than just the data but the entire setup of the project or even access more computing power than would otherwise be available on your own desktop or laptop machine. These form the key elements of Web GIS, which is a large enough and quickly enough growing portion of GIScience that plenty of books have been written about the subject. The Concluding Remarks list some of those. This chapter serves as only a very brief introduction to some of the tools that are available to you through ArcPy.

Just as important as sharing the data or sharing the project is sharing the information you have about the project, which is the project's metadata. Metadata is the information about the data, whether that is your dataset, your project, or a tool you are sharing, and it is critical to the effective implementation of a broader network of GIS practitioners and developers. After all, if all you do is say "here's my data, it's at https://gisportal.myschool.edu/server/rest/services/awesomeData/FeatureServer," nobody will know what it contains, how old it is, what projection it uses, or anything else about the data. To be a valuable contributor in this network, you need to make sure your data is something that other people will want to use. Likewise, putting a .py file online does not encourage people to investigate or use your code, because they won't know what is in it. Even if the code has good comments, they have to be sufficiently interested to download and open the file to even see those comments. The last guided task in this chapter takes you through the addition of some of this metadata to your project.

Task 1: Add Data from on Online Source

The entry point for many people into the world of Web GIS is through a data service. This is accessing a dataset by telling the computer where to go online to find it, rather than downloading the data onto the local hard drive and pointing the computer to that hard drive. This has several advantages. First and foremost among them is that if you are working with a dataset that can be rapidly changing, such as daily meteorological records, or disease counts that are updated on a daily or weekly basis, you do not have to constantly re-download the data every time it is updated to ensure you are working with the most recent version of the dataset. Another advantage is that if the person or organization who created the dataset discovers and fixes an error, by connecting directly to the dataset on their server, you also immediately have access to the corrected version.

From the perspective of ArcPy, there is not much difference between accessing a dataset from the local hard drive versus accessing a dataset from online. In fact, they use the same method of the `map` object: `addDataFromPath()`. The only difference is what goes into the parameter for that function.

© The Author(s), under exclusive license to Springer Nature Switzerland AG 2022
J. Conley, *A Geographer's Guide to Computing Fundamentals*, Springer Textbooks in Earth Sciences, Geography and Environment, https://doi.org/10.1007/978-3-031-08498-0_18

Create a new map, and use the Python Window to create and access a map object called m. Then enter the following line, making sure to have the URL exactly as it appears. This dataset comes from the United States' Health Resources and Services Administration and contains information about areas they have designated as Health Professional Shortage Areas. As the name indicates, they are areas where there are not enough medical providers to meet the needs of the local population:

```
m.addDataFromPath("https://gisportal.hrsa.gov/server/rest/services/Shortage/
HealthProfessionalShortageAreas_FS/FeatureServer")
```

This gives a return value of the Layer object containing that newly added dataset:

```
>>> m.addDataFromPath("https://gisportal.hrsa.gov/server/rest/services /Shortage/
HealthProfessionalShortageAreas_FS/FeatureServer")
<arcpy._mp.Layer object at 0x0000026BA77DE550>
```

This is a group layer with more than one feature layer within it.

Question 1: How many feature layers are within this group layer?
Question 2: How many points are listed within the Mental Health HPSA—points layer?

Task 2: Use a Tool That Accesses an Online Service

The second way many people use elements of Web GIS is through online services. In this approach, instead of accessing data that is otherwise not on your computer, it is running tools that are not stored on your computer. As with accessing data in the first task, this allows the user to access tools that would otherwise be impractical on their own hard drive.

The primary ways this is needed is if the tool requires access to frequently updated information to work properly. This is the case with tools that find drive-time polygons, which is what you will be doing later in this task. Generating a polygon representing all the places within an hour's drive of a specific point or set of points requires the traffic conditions. Of course, traffic conditions are updated very frequently in order to be useful. It is impossible to store all of the relevant information on your desktop computer and maintain it so that it is always current. Instead, the tool is set up as an online service so that it can access that data at the moment it is run, to ensure the most reliable results. The other way that a tool can be beneficial as an online service is if it is too computationally intensive for the hardware and software requirements of most computers. As desktop computers are generally powerful these days, this is not as common for a desktop GIS like ArcGIS Pro. However, for tools that employ algorithms that can benefit from parallel processing beyond the number of processing cores in a standard computer, it can be worthwhile to set it up as a cloud computing service that makes use of more processing cores in a network setting than would be available otherwise. This need for more processing power can also arise in a Web GIS setting if the user is accessing the GIS through a web browser instead of the desktop environment. Web browsers cannot assume or access large amounts of processing power, if only because they are typically designed to be run on a wide variety of hardware platforms, including devices with very limited internal processing capabilities. To avoid overloading the web browser or the computer using that browser, the GIS uses the browser to get input information from the user but handles the computation through a cloud computing setting independent of the browser.

In our ArcPy setting, since most tools are included in the software itself, the first need is more common. Most, if not all, ArcGIS Pro tools that make use of online services require processing credits to run. Therefore, before you continue with this task, make sure you or your organization has enough credits to carry out the task. This can be viewed at https://pro.arcgis.com/en/pro-app/latest/tool-reference/appendices/geoprocessing-tools-that-use-credits.htm, and at the time of writing, the code for this task uses 4.5 credits.

Load the data from Chap. 12, so that you have the states, counties, cities, rivers, and highways visible. We will be finding all the areas that are within an hour's drive time of a city called Springfield. The first stage of this is to create a new layer of just the cities named Springfield. To accomplish this, enter the following code, which should look familiar from previous chapters:

```
arcpy.MakeFeatureLayer_management('cities', 'springfields', "\"AREANAME\" = 'Springfield'")
```

You should then find that there are nine cities in the new layer called `springfields`.

Just as accessing data through an online service looked very similar to accessing data on your hard drive, there is little functional difference between using a regular tool in an ArcGIS Pro toolbox versus using a tool that accesses an online service. In this case, we want the Generate Drive Time Trade Area tool within the Business Analyst extension. Looking at its documentation, just as we would with any other tool, we can see that it requires an input point layer from which the drive time polygons will be based, an output feature class to store the results; what travel mode will be used, such as Driving Time, Walking Time, or Trucking Distance; and the distance that is used for the polygon boundaries, such as 60 minutes or 100 kilometers. In this case, we want to ensure the starting points are the cities named Springfield, giving a sensible output name, and set up a driving time of 60 minutes. This gives the following line:

```
arcpy.ba.GenerateDriveTimeTradeArea('springfields', 'springfields_hour_away', 'Driving Time', 60,
'MINUTES')
```

This might take a while, depending upon how many processing requests the ESRI servers are dealing with at the moment, but you will have a map with drive-time polygons representing an hour from each of the nine Springfields in this dataset. Note that your polygons may look somewhat different from one time to another because of differing traffic conditions. As such, the answer to Question 3 is not guaranteed to be fully consistent from one iteration running this tool to another iteration of the same tool.

Question 3: How many states intersect the Springfield polygons?

Task 3: Create your own online service

Having used services set up by others, you may wish to move on to creating your own service for sharing data that you have. Sharing tools through a Python package is more complex than sharing data and is covered in Chap. 21, so we will focus here on sharing data.

This requires having a connection with your ArcGIS Online account and its associated data portal, so the first thing we will do is sign into that portal. The `arcpy.SignInToPortal` built-in function accomplishes this for us, and it takes as input parameters the site you are signing into (arcgis.com) your ArcGIS Online account username and password. In the example below, I have altered both the username and password:

```
arcpy.SignInToPortal("https://www.arcgis.com", "UsernameRedacted", "PasswordRedacted")
```

As output, you will see a dictionary of information about the account, but this is not something we will use in the remainder of the task. Setting up the data so that it can be uploaded through the service is a process that takes multiple stages. The first of these stages is creating a draft of the service on your local computer. That way, you can make sure everything is as you want before you submit it for online sharing.

This is ultimately through the map's `getWebLayerSharingDraft()` method, but that has enough input parameters that we need to initialize a few things before we can get there. First, as we are creating a draft of the service on the local hard drive, we need to know where it is going. This is set up through an output directory of wherever you want to put this:

```
outdir = "D:\classes\TechIssues\labs\Python\Chap18"
```

There is then the service name, which is what you want your service to be called. As this is an example, I have chosen a name that reflects that purpose:

```
service_name = "FeatureSharingDraftExample"
```

The full file name simply adds the appropriate file extension (`.sddraft`) onto the name of the service:

```
file_name = service_name + ".sddraft"
```

We can then put all of this together for the complete output location for our draft:

```
output = outdir + "\\" + file_name
```

Using four lines and four variables for different portions of the file path might seem like overkill, but some parts of this will be reused for other elements of the process of uploading the draft to the service online. This is, after all, a small-scale version of a strategy we have used throughout this book of breaking a problem down into smaller pieces, so that individual portions can be changed and reused as needed. Now that we have the place in which we are storing our draft, we need to get the contents of that draft set up. For this, we will share a layer of the counties file from Chap. 12.

If you haven't done so already, access the counties layer through the following lines:

```
aprx = arcpy.mp.ArcGISProject("CURRENT")
m = aprx.listMaps()[0]
counties_layer = m.listLayers("USCensus2010_arc")[0]
```

There are different types of servers that we can be using, based upon the data in question. As we want to set up a server to share, or host, vector data, which contains features, we want a server type of `"HOSTING_SERVER"` and a service type of `"FEATURE."` If we were sharing a raster dataset, this would instead be a `"TILE"` service type. The `map` object has a method that allows us to turn it into a service for sharing features, and it will give us the draft we want. Make sure you have the county layer exactly as you want it to appear before running these lines of code, as the symbology will be encoded within this draft and will not be readily changeable afterward. Notice that, while we are using only one layer here, the counties layer, we could include many layers in the list if we wished to share more than just this:

```
sddraft = m.getWebLayerSharingDraft("HOSTING_SERVER", "FEATURE", service_name, [counties_layer])
sddraft.exportToSDDraft(output)
```

You won't likely see any change to your map, so don't panic. If this worked properly for you, then you are ready to turn the draft into a proper service. This has the terminology of a "service definition," as it is still a file located on your hard drive but which contains all the information needed to publish it online. It still abstains from taking that last step of the final publication. This disconnection has a few advantages, especially within larger GIS-using organizations. There can be multiple people involved in the process, as well as many layers being shared. This bifurcation of creating the service from uploading the service allows different people to have those two different roles, as well as allowing many people to create service definitions and place them into a common directory so that all of them can be uploaded at once when everything is prepared.

This illustrates one of the principles of sharing your work. In many cases, it is not a single GIS user sharing their work as an individual but an organization that is creating a GIS portal for sharing larger amounts of data. This means that the practice of sharing the data is only partially about the code for making it happen; it is just as much, if not more, an organizational endeavor that requires cooperation and coordination among a larger group of users within that organization. To take the example from the first task, there is plenty of data about access to health care that is shared by the HRSA. It is unlikely that there is only one person involved in the processes of collecting, maintaining, and sharing that data.[1] Therefore, splitting the two pieces of this process makes sense.

Before we do that, we have to create the file for storing the service definition, which is where we reuse some pieces of creating the draft. We can reuse the file name, stored in the variable `service_name`, as well as the output directory, stored in `outdir`. The difference is that we are creating the service directory itself, with the file extension .sd, instead of the draft with the extension .sddraft:

```
sd_file_name = service_name + ".sd"
sd_output = outdir + "\\" + sd_file_name
```

Now that the name is ready, we can take the first step of that bifurcated process: staging the service. This is done through the following line:

---

[1] I haven't investigated their organizational structure, but am simply making inferences about the amount of work involved in collecting such data and the amount of data they have.

```
arcpy.StageService_server(output, sd_output)
```

Finally, with the service staged and ready to roll out, we can upload it as a hosting server:

```
arcpy.UploadServiceDefinition_server(sd_output, "HOSTING_SERVER")
```

Once this step is completed, you can use a web browser to log in to your ArcGIS Online account and find your shared data. Here is what it shows for me. In this way, data that I have prepared can be shared with the wider world.

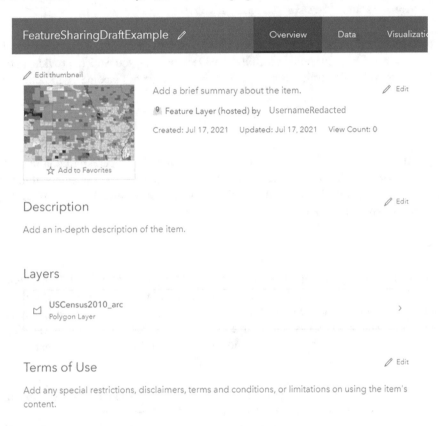

Task 4: Provide Metadata About Your Service

Looking at the image at the end of the previous task, you might notice it is not very useful for anyone who might want to use this service. It even has a generic name that, while descriptive for its purpose within this book, does not convey to a broader, external audience what this service might provide. The description is missing. In a sense, this is like a can of food without its label. There's little way to know what the contents are unless you open the can or access the data. However, who wants to take the time to access data that they know nothing about in the unlikely event it is useful for anything they might want to do?

As mentioned in the chapter introduction, we need to add some metadata to this in order to make it more useful. If you take a look at the different properties of the `FeatureSharingDraft` class, you might notice several of them contain information about the service. By default, these properties do not contain any useful information, but we can edit them to make them more useful.

The first one we will edit is the `.credits` property. The online documentation does not provide as much help as we might want here, as it simply says it is "a string that represents the credits." Since the credits on a map typically provide information about the source of the data, we can provide the source of this dataset:

```
sddraft.credits = "US Census Bureau, 2010 Decennial Census"
```

The next property we will edit is the summary. This, unsurprisingly, provides a brief overview of what the user can expect to see:

```
sddraft.summary = "US Counties, symbolized by per capita income"
```

Third, we can provide a fuller explanation of the contents of the service through the description:

```
sddraft.description = "This feature class shows the per capita income of United States Counties in
2010, using a five-class natural breaks classification."
```

The fourth property we will edit is the `.tags` property. This helps with searching. If you want people who are searching for data to be able to find your service, you should carefully consider what tags you want to provide. In this example, a person searching for "income" might find this service to be useful, so having a tag of "income" will help it appear somewhere in their search results:

```
sddraft.tags = "income, US Counties"
```

The final property that we will edit is the `.useLimitations` property. Depending upon the nature of the data, you might want to specify any recommendations for when you would not use this data. For example, if the precision of the location is coarse, you might want to specify that it is not suitable for use at scales larger than 1:100,000. In this case, we do not have any such limitations, so will specify that:

```
sddraft.useLimitations = "none"
```

Revisiting the documentation for these properties, notice that they are generic strings. The computer, lacking all common sense, will gladly accept any string in here, even "42" or "bad wolf," regardless of how useful or useless it is for the purpose. In that sense, it fills the same function as the comments do within our code. The computer ignores this information and blindly presents it to the user. The difference is that comments are, in most cases, only within a small group of people, perhaps even just for yourself, to help explain the purpose of individual lines of the code. On the other hand, this metadata about

your service is public facing, available to anyone who might come across it. It should, therefore, go without saying that this is an important communication strategy that can help enhance or limit the use of your service.

Returning to the code, we have edited the draft of the service definition. To make these changes public, we have to revisit the code from the previous task turning the draft into a fully uploaded service. There is, though, one more step to do first. In your examination of the properties, you might have noticed one called `.overwriteExistingService`. This records whether or not you are allowed to change the service once it is created. Type in the following line to examine its value:

```
sddraft.overwriteExistingService
```

Question 4: Is the value of this property True or False?

If the value is True, we are ready to move forward with the staging and uploading process. If the value is False, we need to first set it to be True with the following line:

```
sddraft.overwriteExistingService = True
```

Once that is done, we can use the same lines as above to share our updated service with the rest of the world:

```
sddraft.exportToSDDraft(output)
arcpy.StageService_server(output, sd_output)
arcpy.UploadServiceDefinition_server(sd_output, "HOSTING_SERVER")
```

Having completed this, we can go and revisit the online account to check on the changes. We can now see that the summary has been updated, as well as the description. Now that these changes have been made, our shared service will be more useful for those who come across it.

Task 5: Unguided Work

This is a very simple unguided task. Take the ideas from this chapter's guided tasks, and create a new feature service with the layer you used in Chap. 17. Be sure to give it a good title, credits, summary, description, and tags. Provide the URL to find the service.

The evaluation criteria are as follows:

- The service is present at the URL provided.
- The layer from Chap. 17 is included.
- The metadata of title, credits, summary, description, and tags are all present and appropriate.

This chapter will take the ideas of sharing code and use that to incorporate code that has been shared online as a package. This will be carried out using examples from statistical computing using the NumPy package and the arcgisbinding package in R.

Introductory Comments: Python Modules

Sharing a data service and using data that has been shared as a service is, as was stated in the previous chapter, one of the primary entry points for people to join the broader network of GIS users in the Web GIS ecosystem. Data, however, is not the only part of the GIS system that can be shared. The primary focus of this book is developing skills to create scripts to enhance ArcGIS Pro in one way or another. Once you have scripts that are useful to you, it is only logical that the same scripts may be of use to others. The final chapters of this book work toward creating a package that you can share with others.

The first step, though, in creating a package is understanding how they are used. As such, this chapter focuses on the use of external packages that have been created by others. It will use two packages: numpy and arcgisbinding. The numpy package is a statistical computing package and provides a wider array of functions for statistical analysis than those which are automatically included within ArcGIS Pro. If you need even more statistical computing abilities, it may be worth leaving Python entirely for the statistical computing system called R. The arcgisbinding package is in R and allows for communication between ArcGIS Pro and an R programming environment.

There is a wide variety of modules that have been created for Python, most of which have little to do with ArcGIS or spatial data. The concept of modules was introduced in Chap. 5, introducing them as extensions to the base version of Python. By now, you are quite familiar with one of those modules: ArcPy. Just as ArcPy serves a specific extension purpose—allowing the use of Python code within an ArcGIS environment—other modules serve other purposes. Some of these are included with the installation of Python provided through ArcGIS Pro, like the csv module you used in Chap. 5 to read and write CSV files. If you go to the Project tab of the ribbon and click on Python in the menu at left, you will see the Python Package Manager. While this chapter and the associated tasks will not change what packages are available, you can use the Add Packages button to install more packages to the current ArcGIS Pro environment. However, importing packages that are not on this list is not supported, presumably because ESRI will only support packages that they have verified will not cause ArcGIS Pro to crash or otherwise behave strangely.

From a computational perspective, modules are a series of Python files that all serve a common overarching purpose. This means that, if you could investigate the contents of a python module, you could potentially use the strategies from Chap. 5 on reading python code to figure out what each of the components of the module is doing. From a practical perspective, however, there are usually far too many python files to achieve this. The way this multitude of files typically affects what we see or do with the modules is through the error messages. An error message from an earlier chapter is reprinted below. Looking at the Traceback portion of the message is something we have now seen more of, with the tracking of error messages through different functions. In this error message, we see that it refers to an error in File "c:\program files\arcgis\pro\

Supplementary Information: The online version contains supplementary material available at [https://doi.org/10.1007/978-3-031-08498-0_19].

Resources\arcpy\arcpy\arcobjects_base.py," line 90. That python file is one of the many .py files that constitute the `arcpy` module. You can even navigate to that file in Windows Explorer and open it up with IDLE. Should you do so, under absolutely no circumstances would I recommend editing it. That could have a high risk of breaking your installation of ArcPy, and a module as complex as ArcPy is not something you'd want to debug!

```
>>> Ohio_layer.brightness
Traceback (most recent call last):
  File "c:\program files\arcgis\pro\Resources\arcpy\arcpy\arcobjects\_base.py", line 90, in _get
    return convertArcObjectToPythonObject(getattr(self._arc_object, attr_name))
AttributeError

During handling of the above exception, another exception occurred:

Traceback (most recent call last):
  File "<string>", line 1, in <module>
  File "c:\program files\arcgis\pro\Resources\arcpy\arcpy\arcobjects\_base.py", line 96, in _get
    (attr_name, self.__class__.__name__))
NameError: The attribute 'brightness' is not supported on this instance of Layer.
```

Task 1: Turning a Feature Class into a numpy Array

The Python package that we will use in addition to ArcPy is NumPy, a package to support scientific and statistical computing. The examples in the tasks for this chapter come from research I did on comparing measures of access to health care (Conley et al. 2021) and which indeed make use of statistical computing.

One of the key lessons when working with multiple modules is that different modules may use similar, but not identical, data structures. In this case, `numpy` does not inherently use the same data structures as `arcpy`, especially when `numpy` does not have any specific interest in spatial data. This means that we will have to make our first task the conversion of data from an `arcpy`-friendly data structure into a `numpy`-friendly data structure. Conveniently, there is already a function to accomplish this that is provided by `arcpy`.

To make this conversion, we first need the desired data in its `arcpy` format. As we have done before, this starts with getting the feature class. As this will be carried out line by line, the Python Window will be a good place to run the code, although using cells in a Python Notebook would also work:

```
aprx = arcpy.mp.ArcGISProject("CURRENT")
m = aprx.listMaps()[0]
l = m.listLayers()[0]
fc = arcpy.ListFeatureClasses("chapter19")[0]
```

Question 1: Using python code from previous chapters, how many fields are in `fc`?

As mentioned above, this data comes from a project comparing different measures of access to health care. Specifically, it compares a series of measures of socioeconomic and demographic barriers to accessing health care. While many of the variables in the fields of this feature class are used in the calculations of one or more of these measures, they are not needed for this comparison. Instead, only a few of the variables at the end are truly needed. We can set up a list of only the necessary fields through the following line of code:

```
fields = ['HEALTHLINK', 'SHAH', 'BASCUNAN', 'MCGRAIL', 'DOMNICH', 'ASANIN', 'PAEZ', 'GAO', 'YIN', 'DALY', 'CHATEAU_SEFI', 'CHATEAU_SOC', 'CHATEAU_MAT']
```

Lastly for this conversion process, we can access the ArcPy function that will carry it out for us. It is possible, although quite unlikely, that when looking at some of the other functions within the .da module of arcpy in earlier chapters, you noticed this function: arcpy.da.FeatureClassToNumPyArray(). It has two inputs: the feature class (or other table-based data type) that is being converted and the list of fields that are to be included within the numpy array. Enter the following line of code to make the conversion:

```
array = arcpy.da.FeatureClassToNumPyArray(fc, fields)
```

Now that we have our numpy array, you might be wondering why we could not just use a python array as introduced in Chap. 11. Let's take a look at the interior of the array variable here by simply typing its name, array, and pressing enter. You should see output like this:

```
array
>>> array([(0.051485 , 0.07214039, 0.07781311, 0.21964603, 0.26566542, 0.22219777, 0.31320043,
0.21671328, 0.54892157, 0.26455139, -0.08471718, -0.27359168, -0.1790466 ), (0.06597251, 0.06888793,
0.03680725, 0.16410657, 0.27274817, 0.18133654, 0.2993142 , 0.18226777, 0.54615429, 0.2543156 ,
-0.27739867, -0.25147752, -0.35226863), (0.08192977, 0.11589624, 0.0748208 , 0.25928362, 0.30467272,
0.27083206, 0.33296029, 0.20910387, 0.55237061, 0.29681452, -0.08425255, -0.09077854, -0.1229404
), ..., (0.15345213, 0.26639974, 0.40515671, 0.32551684, 0.42989901, 0.29654785, 0.33668167,
0.33963245, 0.88142758, 0.43140419, 0.10041128, -0.38272013, 0.11319308), (0.13256161, 0.23169898,
0.33223293, 0.37609896, 0.50805646, 0.39950759, 0.41706762, 0.3554569 , 0.86723604, 0.41341665,
0.17457706, 0.15773115, 0.23446054), (0.19719181, 0.27864918, 0.41363254, 0.34340663, 0.44583826,
0.32286904, 0.37952758, 0.45799205, 0.8389042 , 0.59461085, 0.30320202, -0.17735393, 0.43767242)],
dtype=[('HEALTHLINK', '<f8'), ('SHAH', '<f8'), ('BASCUNAN', '<f8'), ('MCGRAIL', '<f8'), ('DOMNICH',
'<f8'), ('ASANIN', '<f8'), ('PAEZ', '<f8'), ('GAO', '<f8'), ('YIN', '<f8'), ('DALY', '<f8'),
('CHATEAU_SEFI', '<f8'), ('CHATEAU_SOC', '<f8'), ('CHATEAU_MAT', '<f8')])
```

At first glance, it resembles a somewhat more complicated list, in which it looks like a list of tuples. Note the pattern here of array([(...values...), (...values...), (...values...), ... , (...values...)], ...). Each of the sets of values here, represented by the (...values...) elements of the pattern, constitutes one row in the table. In this case, since there were 13 fields selected, each of these rows has 13 numbers in it. If, like arrays work, all the internal data types are the same (e.g., all integers or all doubles), then this might not seem too different. However, there is another portion after all the values, starting with the dtype name. This, too, seems to be a list of tuples, as it again has the square brackets along the outside, with tuples in parentheses as the values in this list. Instead of one set of values for each feature or row in the table, it has one pair of values for each attribute or column in the table. This means that, in this example, there are 13 tuples, and each of these tuples start with the variable name and then has another value of '<f8' for each variable.

Looking at documentation for NumPy, this is what they call a structured array. Structured arrays allow for more than one data type to be present, although it is not as free form as a list. The dtype list records what type of data is associated with each variable. Once this association is set up, every feature in the structured array must have the same number of attributes, in the same order, using the same data types as specified in the dtype list. In this case, the use of f8 indicates that they are all 64-bit floating-point numbers, which correspond to Doubles in ArcGIS Pro. Even though all the data types in this particular structured array are all the same, this is not required.

Task 2: Using a numpy Function

The means by which these 13 indices are compared is a covariance matrix. While ArcGIS Pro does not easily have a tool to compute such a matrix for vector data[1], numpy does. Therefore, we can use our newly created structured array to find the correlation matrix of all 13 of these indices, telling us the extent to which each possible pair of indices varies together.

---

[1] Strangely enough, it is available for raster data within the BandCollectionStats() function.

You might expect that an `import` statement to be able to access any non-`arcpy` python module within our code. In this case, however, `numpy` is, like `arcpy`, automatically imported when working within the Python Window. This means there is no need for a dedicated `import` statement. Instead, we can just access the `numpy` functions directly, as long as we preface them with the `numpy.` package identifier.

Typing in `numpy.` to the Python Window and waiting will bring up all of the available functions within `numpy`. If you hover over the function in this list, you will get more information about that function. In this case, we will use the function for getting the correlation coefficient of two sets of numbers. This is `numpy.corrcoef()`. Type in `numpy.cor` to be able to access the information about the function we want to use. Hover over that, and look for the answer to the following question. Since the information from hovering does not persist very long, you might need to delete the r at the end of `numpy.cor`, retype that letter, and hover the mouse over the function again to get the answer.

Question 2: What Does the corrcoef() Function Return?

In addition to what it returns, we need to focus on what it uses as its input. While the return value might make it seem like we should be able to give it the array variable as input and get the correlation matrix as the output, this will not work. It, instead, works well with two columns from the table at a time. However, we can use loops to generate a larger correlation matrix.

Before we get to this, let's take a look at a single output. First, note that we access a variable within the array much like we would with getting values from a dictionary. Provide the name of the variable as the index parameter to the array, and it will give us the set of values for that attribute:

```
print(numpy.corrcoef(array[fields[0]], array[fields[1]]))
```

This provides us with a 2 x 2 correlation matrix for the first two variables within the fields list. This matrix is as follows:

```
[[1. 0.56562693]
 [0.56562693 1. ]]
```

It has two values repeated twice. The top left is the correlation between field #0 and field #0. This, of course, is 1, as the values are identical. Likewise, the lower right is the correlation between field #1 and field #1. These are not really useful for us, although they are part of the correlation matrix. The other two values are the correlations between field #0 and field #1 in the top right and between field #1 and field #0 in the lower left. Since correlations are symmetrical, these two values are the same.

To generate a full correlation matrix for all 13 variables, we need to loop through each pair of values. To help us plan, let's start with an outline and recognize that we want our correlation matrix to be a list of lists. I have set this up as a series of comments:

```
# First initialize the correlation matrix as an empty list, and an index variable to 0.
# Loop through all of the fields.
# Start a new empty list to be the next row in the correlation matrix.
# Loop through all of the fields again.
# Get the correlation coefficient for the current pair of fields and add it to the end of the row.
# Add one to the index at the end of the row so we can start the next row of the correlation matrix.
```

Filling in this set of comments, we start with the initialization step.
First initialize the correlation matrix as an empty list, and an index variable to 0:

```
corr_matrix = []
index = 0
```

Looping through all of the fields suggests the use of a `for` loop, going through each field.
Loop through all of the fields:

```
for i in fields:
```

Now we need to initialize the row within the matrix. Note that the matrix itself was initialized only to an empty list. This means that since each element within the list is itself a list, we also need to initialize this interior list as an empty list, and add it to the end of the correlation matrix so that it is guaranteed to start a new row.

Start a new empty list to be the next row in the correlation matrix:

```
→ corr_matrix.append([])
```

To populate each element within this row, we need to again loop through the matrix. The nested `for` loops ensures we compare every possible pair of fields. Even though the two loops are using the same list, they have different index variables, marking different locations, so one will not confuse the other. Imagine having two cursors looking at the same table, CursorA and CursorB. The position of CursorA is not altered by CursorB, nor is the position of CursorB affected by CursorA. They can cycle through the same table independently of each other. The same dynamic can apply here, ensuring i and j do not interfere with each other.

Loop through all of the fields again:

```
→ for j in fields:
```

Now that we are looking at a single pair of fields—field i and field j—we can get the correlation between these two and add it to the end of the list representing the current row inside the matrix. The first line below gets the correlation coefficient matrix between the two variables, but we only want one of the more useful values. The index pair `[0][1]` indicates item #1 (the second item) in the top row, giving us the top right value. Getting the lower left value would be just as useful, but this is the pair I chose to extract. Once it is available, it can be appended to the end of the current row in the correlation matrix. That is referenced through the variable called `index`, which we initialized at 0 so it starts by creating the first row.

Get the correlation coefficient for the current pair of fields and add it to the end of the row:

```
→ → corr = numpy.corrcoef(array[i], array[j])[0][1]
→ → corr_matrix[index].append(corr)
```

At the end of the row, we then need to increment the variable `index` by one, so that when we start the next row, it is going to be looking at the correct spot in the matrix.

Add one to the index at the end of the row so we can start the next row of the correlation matrix:

```
→ index = index + 1
```

Putting this all together gives the following code block:

```
corr_matrix = []
index = 0
for i in fields:
→ corr_matrix.append([])
→ for j in fields:
→ → corr = numpy.corrcoef(array[i], array[j])[0][1]
→ → corr_matrix[index].append(corr)
→ index = index + 1
```

Question 3: Using the indices in corr_matrix, what is the correlation between the MCGRAIL and ASANIN indices?

Task 3: Preparation for Connecting ArcGIS with R

The other package we will look at is the `arcgisbinding` package in R. This set of tasks will illustrate two lessons. First is connecting another platform with ArcGIS Pro through the ability to run R scripts in a newly created ArcGIS Pro tool. The other lesson is an introduction to Github, which is one of the primary venues where code is shared. We have to go to Github before we can go any further in this set of tasks because the connection between R and ArcGIS Pro is not something that is automatically installed with ArcGIS Pro. Instead, it must be downloaded and installed separately.

Before we get to the details here, I'll give a brief introduction to Github. It is one of the primary online repositories of open-source code. Open-source software represents a different mentality from a lot of software, such that for open-source software, any person can download the code for that software and is able to make changes to it. This means that for the r-bridge-install project that we will use, you can download and examine the Python code that makes it up. (Github accepts code in other languages as well; it just so happens that this one is in Python.) By contrast, there is no way to examine the source code for ArcGIS Pro. You can see at least parts of `arcpy`, but that is different from the code that Windows uses to run ArcGIS Pro.

To install the connection between ArcGIS and R, first go to the following URL to access the correct software within the Github repository: https://github.com/R-ArcGIS/r-bridge-install. This website might be a bit intimidating when you first open it up. Most of what you see at the top of the page are items that are useful for those who need to see the exact scripts for this project, whether they want to just examine them or edit them. The row of buttons at the top provides different functions for those who are contributors and heavy users of the software in this project. We won't need to look at these. As long as the top row is still actively on the button labeled "<> Code," you can see under that row a list of folders and files. This is the code itself and a few associated files. The main thing to observe here is the set of ages at right. That represents the most recent edit to that file or to any file within the folder. As of the time of writing, only two of these are less than a year old, and most are considerably older. This can be a good sign as it indicates stability, assuming the code is routinely used. In this case, it is. Thus, for the code to be used frequently, but not edited frequently, it must be working very well—well enough to not be worried about finding many, if any, bugs.

We are more interested here in the next section below that, which is labeled "Install the R-ArcGIS Bridge." The first thing to notice here is that there are a couple different installation versions for the R-ArcGIS Bridge. Since we are working with ArcGIS Pro, we want the 64-bit installation of the R-ArcGIS Bridge to ensure it is compatible with ArcGIS Pro. Scroll down to the area labeled "ArcGIS Pro 1.1+" and the "Offline Installation" below that. While either the online or offline installation would work, the offline installation process will probably be a bit more robust in case of internet connectivity difficulties, or simply working on a laptop that you want to travel with. Download the files through the "download this repository" link. Then click on the "latest version of the arcgisbinding package" link, and select the file named arcgisbinding_1.0.1.244.zip[2]. Save both of those, and copy to a directory on your hard drive. The exact address of that directory is not important, although I do recommend something that is sensible and easier to remember because you'll need that directory once we switch over to ArcGIS Pro to make the connection.

Because the files downloaded in the .zip format, we need to extract the contents of those .zip files before ArcGIS Pro can use them. Navigate to the directory you just saved the files to, and extract them into that directory. Then, open up ArcGIS Pro, and go to the Catalog window. If you do not see it, find the View ribbon and click the "Catalog Pane" button. Then find the Toolboxes menu item, and right-click on that item. In the menu that pops up, select "Add Toolbox," and navigate to the toolbox you just extracted from the first downloaded zip file. It is in the r-bridge-install-master folder and is called "R Integration.pyt."

Within that toolbox, you will find a tool called "Install R bindings." Run that tool. You will probably get a "completed with warnings" alert, but that's no need to panic. At this point, the bridge is set up for connecting ArcGIS Pro and R.

Task 4: Creating a Tool in ArcGIS to Use an R Script

Now that it is installed, we can create a tool in ArcGIS Pro using the same steps that we used to create a tool in Chap. 14. The difference is that whereas in Chap. 14, we used a Python script that we created, in this task, we are using an R script that is provided for us. This tool carries out a type of regression called spatial lag regression. This particular statistical technique

---

[2]That's the version number as of the time of writing. This can, of course, get updated whenever they feel like it, so don't panic if the version number is different.

is not (yet) installed with ArcGIS Pro by default, so if we want to run it, a new tool must be created. There are packages in R that have a much wider variety of statistical techniques than are available in ArcGIS Pro. Therefore, a comparatively straightforward way to be able to use these additional statistical techniques within the ArcGIS Pro environment is to be able to take an R script that carries out the test and be able to run it within ArcGIS Pro, now that we have a bridge between the two.

First, create a new tool as you did before. To recognize what it is, use the Name "SpatialLag," and a Label of "Spatial Lag Regression." For the script file, navigate to the provided file of spatial_lag_test.R.

We will then need to provide parameters for this new tool. It needs to have the feature layer that is being analyzed as the input.[3] This provides the data for the analysis. We also need to choose the dependent and independent variables, as a spatial lag regression is set up like any other regression.[4] As the second input parameter, use a single field as the dependent variable. We will also want to make sure this field is in the dataset, so it must have a dependency on the input dataset. The independent variables will also be set up as an input variable with Field as the data type, except when you pick the Data Type, make sure the box labeled "Multiple values" is checked. It also has a dependency upon the input dataset to ensure the independent variables are contained in this dataset.

Now we can run this tool. Try it with the chapter19 feature class as the input data, privbeh30 as the dependent variable, and HEALTHLINK, SHAH, and BASCUNAN as independent variables. This runs a spatial lag regression in which the dependent variable is the number of private behavioral health facilities within 30 minutes of the centroid of the block group, and the three independent variables are three different multivariate measures of socioeconomic barriers to health-care access. Click Run.

It should say that it completed with warnings. If not, you might have to save your work, close ArcGIS Pro, then go back to the Windows start menu, and right-click on ArcGIS Pro to run it as an administrator. (If this happens, I hope you have administrator access on your computer!) Once you have it running as an administrator, redo the Install R bindings tool, and re-add the tool for this task.[5]

Once it ran, you might get a very large number of warnings, but they did not seem to impede the ability of the script to complete, as it printed out the results at the end of all the messages. Within the spatial lag regression results, one of the main pieces of information we are interested in is the value for Rho, which is presented under a table with the independent variables listed. Because the focus of this is running the script, not explaining the statistics, you won't be expected to provide a statistical interpretation of Rho here.

Question 4: What is the value of Rho in this regression?

Task 5: Examining the R Script

This task takes a look at the R script that was attached to a tool in the previous task. The purpose here is not to make you an expert in programming R; learning Python is clearly enough for a single book without having to add any other languages! It is, instead, here to illustrate some of the similarities between the two languages and thereby demonstrate an argument from the first section of the book. Once you learn the fundamentals of programming and can recognize the syntax patterns of different general strategies, it becomes considerably easier to decipher code from another language and eventually program in it. The script is below:

```
#Spatial Lag Regression R Script
tool_exec<- function(in_params, out_params){

 arc.progress_label("Loading packages...")
 arc.progress_pos(20)
```

---

[3] I figured this out by looking at the R code, seeing that this input parameter will be used as an input parameter to a command called arc.open, and looking at the documentation for arc.open to identify what it expects for its input parameter.

[4] Strictly speaking, it should also ask for a spatial weights matrix, but that is probably more work than is needed at this point in time, so we will assume a queen contiguity matrix.

[5] I tested this on two machines. I had to "run as administrator" on one, but not the other, and I'm not sure what the difference was. On the machine that required me to run it as an administrator, there were fewer warnings, while the one that didn't require that step gave many more warnings.

```
if(!requireNamespace("spatialreg", quietly = TRUE))
    install.packages("spatialreg", quiet = TRUE)
if(!requireNamespace("spdep", quietly = TRUE))
    install.packages("spdep", quiet = TRUE)
install.packages("spatialreg", quiet = TRUE)
require(spatialreg)
require(spdep)

input_data <- in_params[[1]]
dependent_variable <- in_params[[2]]
independent_variables <- in_params[[3]]

arc.progress_label("Loading data...")
arc.progress_pos(40)

d <- arc.open(input_data)
fields_list <- append(c(dependent_variable), independent_variables)
d_df_full <- arc.select(d)
d_df <- arc.select(d, fields = fields_list)

d_sp <- arc.data2sp(d_df)


arc.progress_label("Running Regression...")
arc.progress_pos(60)

response <- d_df[, 1]
predictors <- d_df[, -1]

queen <- poly2nb(d_sp, queen = TRUE)
queen_nb <- nb2listw(queen)
lagModel <- lagsarlm(response ~ ., data = predictors, queen_nb)

arc.progress_label("Writing output...")
arc.progress_pos(80)

if(!is.null(lagModel) && lagModel != "NA")
    print(summary(lagModel))

arc.progress_pos(100)
}
```

The first thing to notice is that it is entirely within a single syntax pattern as below. The word `function` there can give a clue as to what is happening. It is a keyword for setting up a new function, just like the `def` statement in Python. This, then, has two input parameters for the function, `in_params` and `out_params`. In place of the colon and indentation to indicate the beginning and end of the code block within the function, R is using curly braces at the start and at the end:

```
tool_exec<- function(in_params, out_params){
    ...
}
```

This would then be equivalent to the following Python line:

```
def tool_exec (in_params, out_params):
```

You might observe, though, that there is no line in this file that actually executes the function. That hints at one of the more computationally complex aspects of this bridge. All examples consist solely of a function named `tool_exec`, indicating that when the tool is run, the bridge is looking for a function by the name of `tool_exec` , and automatically runs that function.

The next line after the function definition, `arc.progress_label("Loading packages...")`, looks like it could be accessing the method of an object. This is similar to what is going on, by setting the label that is attached to the progress bar in ArcGIS Pro. The difference is that dots are permitted in variable names and function names within R, so this is simply a function called `arc.progress_label`, not a method of some variable called `arc`. Looking at the PDF documentation for the `arcgisbinding` project[6] will show that almost all of the functions provided by the package begin with `arc.` including the dot. This illustrates that when comparing one programming language against another, the delimiters and keywords are often what can change the most.

After the two functions for updating the progress bar are run, we see another mostly familiar type of structure: the conditional statement. The same keyword of `if` is used, and in both cases, a Boolean expression follows that keyword. While they were not needed here, the curly braces could have been used again in place of the colon, like they are used for the function definition statement. The lines inside the conditional statement perhaps provide a better clue to what it is doing than the condition itself, at least for those who are unfamiliar with R. The function name `install.packages` suggests a purpose similar to importing a module, and that is, indeed, what this is doing. The conditional statement is making sure a module is imported only if it is needed—i.e., it hasn't already been loaded. A fairly quick search about the require function reveals that it is what R uses to import a package within a function, as we are in this case. It also shows that for R packages, importing a package is a two-step process, where it first must be installed, then it must be accessed through the `require` function or the `library` function if you are not in another function.

The next three lines all seem to be accessing values from the input parameters, and the use of variable names corresponding to the input parameters from the tool makes this very likely. It shows the use of R's assignment operator. Where Python uses a single equals sign (=) to make an assignment, R uses a two-character left-facing arrow (<-).

The two lines after accessing the input parameters are, again, updating the progress bar's label and position. The subsequent five lines are, at first glance, fairly straightforward assignment operators. If you take a look at the `arcgisbinding` documentation, you might begin to decipher the different functions that are used here. The first one, `arc.open`, as the name indicates, opens up the input data layer. Then the two sets of variables are appended to get a single list of fields that needs to be retrieved from the dataset. This retrieval is in the `arc.select` function. The last of these lines, `arc.data2sp`, converts the data into a spatial data frame, which is a data structure specific to R. This reinforces the issues of working with `numpy`, in that transferring from one package to another, or in this case, one language to another, frequently requires the conversion of data from a structure suitable for the first package to one suitable for the second package.

As you look through the remainder of the script, observe that most of the lines are variations on the ones already explained here, whether that is setting the label or position of the progress bar, assigning values to variables, or using a conditional statement to control the execution of the code. You will probably even notice the `print` function looking very similar.

This task should help provide confidence that, upon learning the basic fundamentals of computing, such as variables, assignments, control statements, and functions, it becomes easier to switch among different programming languages. Therefore, through the skills contained in this book, you should have not only a guide to computing with Python in ArcGIS Pro but really a guide to GIS computing in general for when—not if, but when—ESRI decides to change from Python to another language, or you decide to move to a different GIS platform, like QGIS.

Task 6: Unguided Work

We may wish to have the data for a two-dimensional histogram, since the Charts function within ArcGIS Pro only supports the creation of a one-dimensional histogram. Displaying the histogram itself would not work within the Python Window, but we can still get the number of block groups that would appear in each bin.

---

[6]Currently at https://r.esri.com/assets/arcgisbinding.pdf

Write the code to generate the bin counts for a 2D histogram of the percent of residents with no health insurance (*pctnoin-sto*) and the unemployment rate (*unempRate*) with ten bins in each of the two variables.

You will want to use the `numpy.histogram2d` function, which has inputs similar to the `corrcoef` function used earlier.

It will provide output similar to this, although I used a different set of input variables. The top part contains the number of features within each bin. The second and third parts of this output are the breakpoints used to generate the bins of the two input variables. For example, the lowest bin in the first variable contains all features less than 0.07734367, and the lowest bin in the second variable is all features with values less than 0.10930839. There are six features with the lowest class in both variables.

```
(array([[ 6.,  0.,  0.,  0.,  0.,  0.,  0.,  0.,  0.,  0.],
 [ 1.,  2.,  3.,  0.,  0.,  0.,  0.,  0.,  0.,  0.],
 [ 0., 18., 30., 13.,  1.,  0.,  0.,  0.,  0.,  0.],
 [ 0.,  3., 88., 160., 52., 11.,  3.,  0.,  0.,  0.],
 [ 0.,  0., 14., 168., 249., 84., 19.,  4.,  1.,  0.],
 [ 0.,  0.,  2., 31., 138., 191., 43.,  7.,  3.,  0.],
 [ 0.,  0.,  0.,  3., 28., 63., 59., 17.,  2.,  3.],
 [ 0.,  0.,  0.,  0.,  2.,  9., 10., 15.,  6.,  2.],
 [ 0.,  0.,  0.,  1.,  0.,  0.,  3.,  8.,  9.,  3.],
 [ 0.,  0.,  0.,  0.,  1.,  0.,  0.,  2.,  0.,  1.]]),
 array([0.07734367, 0.12433297, 0.17132227, 0.21831157, 0.26530087, 0.31229017, 0.35927947,
0.40626878, 0.45325808, 0.50024738, 0.54723668]),
 array([0.10930839, 0.15922085, 0.20913331, 0.25904576, 0.30895822, 0.35887068, 0.40878314,
0.4586956 , 0.50860805, 0.55852051, 0.60843297]))
```

Evaluation criteria are as follows:

- The code provided works within the Python Window or Notebook to produce the `numpy` data structure(s) necessary for the histogram function.
- The code provided correctly employs the `histogram2d` function.
- The output generated is correct.

Introductory Comments: More Complex Programming and UML

The last part of sharing your work that we will cover in this book is creating a Python package. The details of that are in the next chapter, but, first, we will look more at the programming process that often leads up to the creation of the package. In many situations, whether for creating a package or not, you may find yourself in a position where you want to be using more than one Python file. It may be a situation where a single file is becoming too long to manage, so you want to break it up. More common, though, is a situation in which you realize some parts of the code you are writing could be more generically useful and would be beneficial to be a separate file that is more readily accessible to a wider range of projects.

This is the case in the situation that inspired the code for this chapter. I was working on a research project in which I created hundreds of feature classes of simulated data for statistical analysis. This thus far unpublished project looked at the sensitivity of statistical p-values to certain geographic situations, and I assessed this sensitivity by comparing the results of LISA analysis of the US 2016 presidential election against LISA analyses of randomly simulated data. There were 100 simulations for each LISA analysis, each of those simulations being in its own feature class. For part of this work, I wanted to create a list of the values for each county across all 100 simulations and find the minimum, maximum, mean, standard deviation, and other analytical summaries for this list of numbers. The standard approach for GIS would have been to join all the simulation feature classes, but joining 100 feature classes seemed like overkill for this situation. While I could perhaps have had 100 cursors stepping simultaneously through the feature classes, there isn't any guarantee that they will have all features in the same order, which is necessary for the summaries I was computing.

Therefore, instead, I decided to create a class called a multiCursor that can act like a cursor in stepping through the files, but where a cursor returns a list for the single row corresponding to that feature's entry in the attribute table for the feature class associated with that cursor, in the multiCursor, it returns a list of lists, with each internal list corresponding to that feature's entry in one of the feature classes and the overall list containing one list for each of the feature classes provided to it. It would also be useful if the multiCursor also has methods to support finding all values associated with a specific feature, and to support finding all values associated with a specific field.

This concept of a multiCursor can be useful in additional situations, not just this specific p-value investigation, so I wanted it to be a separate Python file that can be accessed. Likewise, I felt that the function carrying out the summary statistics could also be more broadly useful, so set that out as its own file. Then there is a third file for this specific research project.

As this project becomes more complex, there should be a way to organize the structure and represent it, to ensure that all parts get completed, and that it is easier to identify the relevant parts of the code for later communication about it. This way is using a Unified Modeling Language, or UML. It was briefly introduced in Chap. 12 to show the connections between different classes involved with the mapping module, and a bit more is added here. UML is a large and very structured approach to creating diagrams to assist in software development. Since UML was designed for much larger projects than this, a full implementation of UML is probably overkill. Even so, the use of class diagrams can still be useful and can help illustrate how we can organize our thoughts before attempting to write the code.

Supplementary Information: The online version contains supplementary material available at [https://doi.org/10.1007/978-3-031-08498-0_20].

More than anything else, I hope that what you get from this chapter is a sense of the planning that is involved in a larger programming project. To return to the basics, we break down a big project into smaller and smaller sub-projects until those sub-projects become manageable chunks of code, and then, if the planning has gone well, they can easily slot into place for the bigger picture, and you have a fully functioning multiple-file program. Even if you do not use UML or anything like it, you will almost certainly require much more planning to execute a larger project than anything that you did for the previous chapters.

Task 1: Construction of UML-Like Diagrams

Building upon the writing analogy from earlier chapter, it can be daunting to just stare at an empty file for a large task like this, just as it can be daunting to stare at an empty page for a longer essay. Where you might use an outline to organize your thoughts prior to writing an essay, programming might not be as strictly linear, so an outline of a complex project is perhaps not helpful. Instead, a class diagram shows the relationships between classes in a programming project that typically contains many different classes. Since class diagrams are designed with object-oriented programming in mind, in which everything is an object and all programming exists within classes, and we do not have a fully object-oriented system here, we will have to adapt the class diagram somewhat to accommodate the parts of the program that use a function-based style of programming instead.

I start here with the main analytical program that we will attach to an ArcGIS Pro tool in Chap. 21.[1] It has the general code that isn't part of any function, primarily to import the parameters, and a function to carry out the analyses. In the middle block of the class diagram[2] element below, I am denoting the input parameters with their names and their types, as they will be set up through the custom tool in Chap. 21. In the lower block, the function is specified. The first items under the name of the function are the input parameters. These will go into the tuple of parameters passed into the function, and the diagram also specifies the data types they will be. Lastly, it notes that there is no return value. We will see why when we fill out the plan for this function in Task 2. Also with respect to that plan, you will notice that this diagram specifies nothing about the code itself or what the contents will be. It treats it simply as a black box, specifying the inputs and outputs, and nothing else (Fig. 20.1).

The next file will help out with the computation of the summaries. Its purpose is to take a list of numbers or test values, along with the procedure that is being used. Unlike the first file, it takes no information from the user and does not create any class with an `__init__` function. This means the middle block of the UML element is empty. The function is still presented in the lower block. It specifies what the input parameters are. Following the lead of the arcpy documentation that we have seen, I also indicated optional input parameters by placing their names in curly braces. Unlike the `populationSummaries` function, this function has a return value, which will be a floating-point number (Fig. 20.2).

The third file is the one best suited for the UML diagram, as it contains a class: the `multiCursor`. All the file contains is the class, so we will only have the class itself shown. The middle component of the diagram element has the parameters

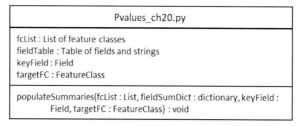

Pvalues_ch20.py
fcList : List of feature classes fieldTable : Table of fields and strings keyField : Field targetFC : FeatureClass
populateSummaries(fcList : List, fieldSumDict : dictionary, keyField : Field, targetFC : FeatureClass) : void

Fig. 20.1   Diagram of the PValues_ch20.py file

---

[1] We aren't doing this now because, for some reason, it is much easier to debug the program using IDLE. The tool will read the connected files, but if there is an error in a connected file that you need to debug, somehow, that fix doesn't readily show up. It is as if the connected files were compiled and stored as the compiled binary files instead of something that can be changed later.

[2] This is, strictly speaking, not exactly a class diagram, because it is referring to a file instead of a class. This is because the file here uses functional programming instead of object-oriented programming, and UML is designed only for object-oriented programming. The use of internal variables set from the parameters of the tool as similar to properties, and a function in place of an object's method(s) is an adaptation of UML that, in my opinion, helps to explain what is going on in this file. The same is true of Fig. 20.2.

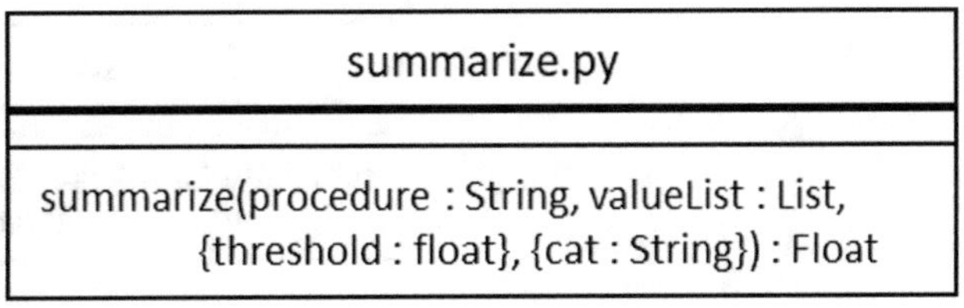

```
┌─────────────────────────────────────────────────┐
│                  summarize.py                   │
├─────────────────────────────────────────────────┤
│                                                 │
│  summarize(procedure : String, valueList : List,│
│       {threshold : float}, {cat : String}) : Float│
│                                                 │
└─────────────────────────────────────────────────┘
```

Fig. 20.2 Diagram of the summarize.py file

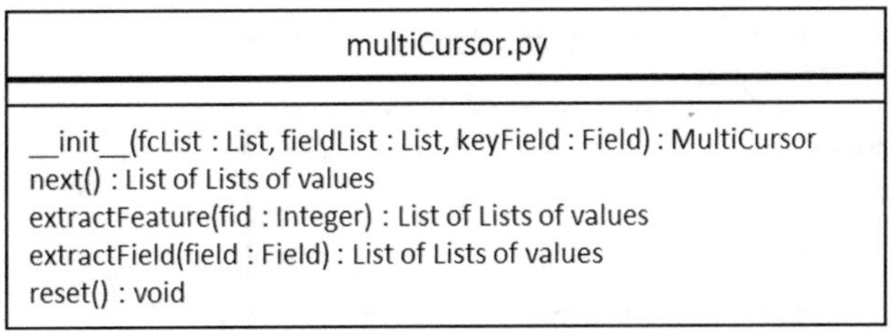

```
┌─────────────────────────────────────────────────────────────┐
│                       multiCursor.py                        │
├─────────────────────────────────────────────────────────────┤
│  __init__(fcList : List, fieldList : List, keyField : Field) : MultiCursor │
│  next() : List of Lists of values                           │
│  extractFeature(fid : Integer) : List of Lists of values    │
│  extractField(field : Field) : List of Lists of values      │
│  reset() : void                                             │
└─────────────────────────────────────────────────────────────┘
```

Fig. 20.3 Diagram of the multiCursor.py file

used to create an object of this class. They will be the inputs for the `__init__` method (except for the obligatory `self` parameter). There are four methods. The first is like the `next()` method for a cursor. It has no input but provides the output for the next feature. Unlike the cursor, which returns a list of values, this returns a list of lists of values, with each internal row corresponding to the next feature's values for one of the input feature classes. The `extractFeature()` method functions much the same way but allows the method to specify the feature of interest, rather than simply get the next one. The `extractField` function is for when you are interested in all values across all features for a single attribute. It returns a list of lists of values, like `next()` or `extractFeatures()`, but whereas in those two methods, each row corresponds to an input feature class and each column corresponds to an attribute, here, each row corresponds to a feature, and each column is the value of that field for that feature in all of the input feature classes. The last is the reset method, which like the reset method of a regular cursor, sets the cursor back to where it is pointing at the first feature (Fig. 20.3).

There is value in having these diagrams that provide the bare-bones structure of each file or each class if you have more than one class in a file. The bigger value from UML, though, is from representing how they are connected, which is what we used them for in Chaps. 12 and 13. Again, because of the somewhat off-label application of UML to code without any objects, this is an imperfect representation of UML but can be helpful for organizing the code. The addition of relationships between the files is added below, to make clear that PValues_ch20.py will call summarize.py and will use exactly one multiCursor (Fig 20.4).

These relationships have arrows, which also is useful in making sure we know the dependencies in our files are going to go the correct direction. With this done, we can then begin to plan out the components of the three files.

Task 2: Plan the Components of the Files

I'll take the three files in the same order and plan them out. The approach here is parallel to the illustrative example at the start of Chap. 17. This section creates the flow charts and pseudocode for each of the three files, and the next three tasks fill in the pseudocode. I will skip the debugging aspect of Chap. 17, however, because it would simply take too long! I also note here that the presentation of this is a more linear process than what actually took place. There was one case where I realized that, in fact, a different method would be very useful and went back to the UML concept, added a method, and returned to planning the code out. This will be explained when I get to that part of the relevant flow chart. Likewise, I weighed different options for some of the control statements, tried planning out several approaches, and only present the one that I ultimately settled on below. Again, a brief explanation is given at the appropriate spot.

Just like we started out the flow chart for Chap. 17 very simply, the flow chart here for the PValues_ch20.py file starts out very simply (Fig 20.5).

Using the UML diagram above, we can recall what those input parameters are going to be a list of feature classes, a table of fields and strings, another field, and a feature class. The data types do not tell us very much, but here with our plan, we can

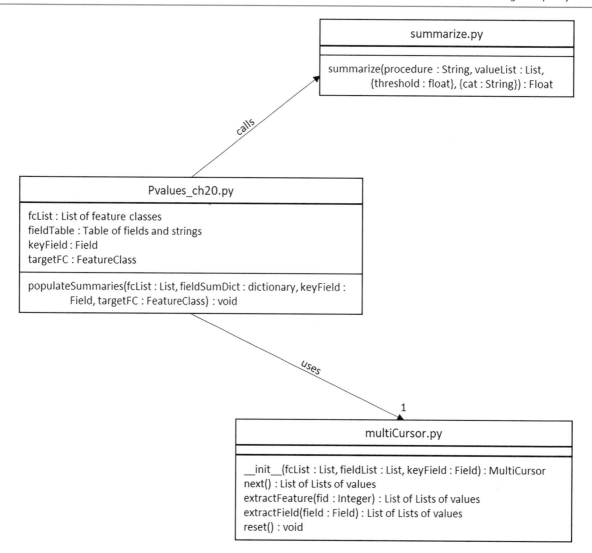

Fig. 20.4   Relationships among the files

Fig. 20.5   Beginning flow chart for PValues_ch20.py file

fill out some of the intent behind the parameters. What we need from the user starts with a set of feature collections. These are the simulations in the motivating application from the introductory comments, although it can be any list of feature collections. We also need a table of fields and strings. This is an input type for ArcGIS Pro tools that I have not yet covered. It allows for input parameters to be entered in pairs (or groups of more than two, but pairs here). This way, the user can specify each pair of a field to be summarized alongside how it is to be summarized, such as finding the minimum value, the maximum value, the mean, etc. The field is, of course, that field, and the string is the summarization procedure. It should be noted that all of the fields in this table must be present in all of the feature classes in the fcList input parameter. The keyField is a unique identifier that will be used to keep track of the features as we cycle through them. Lastly, there is another feature class that will store the output of all the summaries in new fields. This target feature class does not need to be one of the input feature classes in fcList.

Fig. 20.6 Filling in the black box for PValues_ch20.py

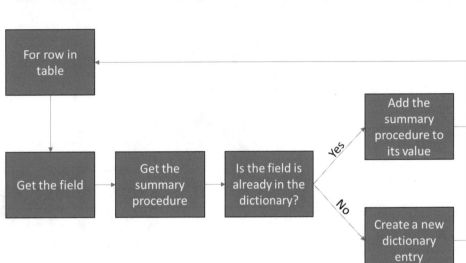

Fig. 20.7 Converting a table to a dictionary for PValues_ch20.py

The black box in the flow chart above will primarily be the populateSummaries function, but we can compare the parameters that we get from the user with the input parameters for the function and notice that there is one change. The function takes a dictionary, rather than a table. This means we need to somehow convert the table into a dictionary before we can run that function (Fig. 20.6).

The means by which we carry out that conversion might not be straightforward, so let's flesh that out. The first step in this is to figure out what we want this dictionary to look like. For creating the summaries, it would probably be useful if we had a record of all the summaries requested for each field. This lends itself to a dictionary in which the field is the key and the value is a list of all the summary procedures for this field. The available summary procedures are minimum, maximum, mean, standard deviation, counting the number of features that are less than a value, and counting the number of feature that are equal to a value. The user, though, can request as many or as few of these as desired (Fig. 20.7).

Once the dictionary is created, we can move to the box labeled "Run populateSummaries function" in the flow chart and thereby run the function that uses the dictionary and the other user-provided parameters, to calculate the summaries and put them in the right attributes of the target feature class. This last box in the overall flow chart, which is "Put summaries in target FC," will be part of the function, and since that is the goal of the function, this is why there is no return value for the function. Instead of returning a value to the Python code, it instead puts those values in the feature class designated for that purpose.

This is a more complicated flow chart, because it ultimately requires running the populateSummaries function for each combination of feature, field, and summary procedure. This means we are going to need a trio of nested for loops. I had initially considered looping through each field first, then each summary procedure, and each feature within that, but it became challenging to figure out how to correctly manage the update cursor. Because of that, I concluded that the simplest approach was to tackle each feature one by one, instead of each field, and therefore have the outermost loop be through the features (Fig. 20.8).

Within the feature loop, I first extract all the values for that feature from the multiCursor. Once those values are extracted, I have a list of lists of the values. However, to get the summaries, I need a list of just those values for the field being examined at each stage of the "for each field" loop. Once I have that single list, I can give that list of values to the summarize function, along with the desired procedures, one by one in the innermost loop. Summarize gives me a single value, which I can place in the correct position in the update cursor. When all field and procedure combinations are completed for the feature, that row of the update cursor must be updated before moving on to the next feature.

In this flow chart, we use both the summarize and multiCursor files. However, as far as this file is concerned, we are trusting that they do their work correctly and will not expand upon those boxes in the flow chart here. By splitting this up, we can separate out the complexity of computing the summaries and keeping track of the features and fields and feature classes

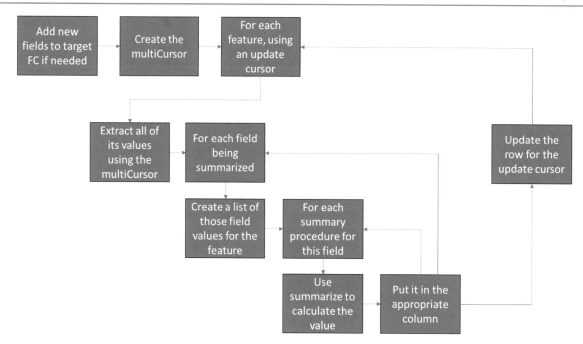

Fig. 20.8 Calculating all the summaries within PValues_ch20.py

within the multiCursor. Putting the entire flow chart together, we get the following. As you can see, just this single file takes up the majority of the page, again reinforcing the idea that splitting the overall task up into three files makes it more manageable (Fig. 20.9).

Next, we turn to the summarize.py file. This only has one function, and the UML diagram represented its inputs and output. That informs the beginning of this flow chart (Fig. 20.10).

This is still not very helpful, so let's fill in the black box. The main thing is that it takes a list and a procedure. There are the optional parameters as well. What happens inside the black box is dependent upon the procedure that is chosen, so it makes sense to use a conditional statement here, selecting from among the different possible procedures, calculating the right one, and returning the appropriate summary value. Notice that the optional parameters of threshold and category appear for certain procedures only. This is why they are optional. There's also an alert for the situation in which the stated procedure does not match any of the options (Fig. 20.11).

Once we have this, we can slot this into the black box of the entire summarize.py flow chart and get the following (Fig. 20.12).

Lastly, we can move on to the flow charts for the multiCursor.py file. There will be one per method. Conveniently, at least, some of these will be simple flow charts. We will take them in the order in which they appear in the UML class diagram. The first one is the `__init__` method, implied by the constructor in the middle box. The main purpose here is to create properties of the object's `self` that correspond to the input parameters. Not shown in the UML diagram, though, are the two additional properties that are used to keep track of the internal state. They are the cursor associated with the first feature class and the index of the key field within the list of fields. The purpose of both of these will become clear in later flow charts (Fig. 20.13).

The first method listed in the UML diagram is the `next()` method. It has no inputs but can access the properties, including the newly created ones. It must return a list of lists of values, and these values are associated with the next feature in the multiCursor. This is where the cursor for the first feature class comes in handy. We will simply advance that cursor to the next feature and give that feature to the `extractFeature` method (Fig. 20.14).

This means the real work is done by the next method: `extractFeature`, which has an additional input of the feature's ID value within the key field. It starts by setting up the output as an empty list that we can append things to later. We will also want to have a query that we keep using to isolate that feature's values within each feature class. Using that query, we can loop through the list of feature classes, which is stored as a property for this class, get a search cursor for that feature class, and use it to extract the row corresponding to the feature we are interested in. That row, which is represented as a list of values, can itself be appended to the output, thereby creating the desired list of lists. This is the method for which I had to revisit the UML scheme. Initially, I had planned to not have a separate `extractFeature` method and be able to use the next method to get the values, but I realized while creating the flow chart for `populateSummaries` in the PValue_ch20.

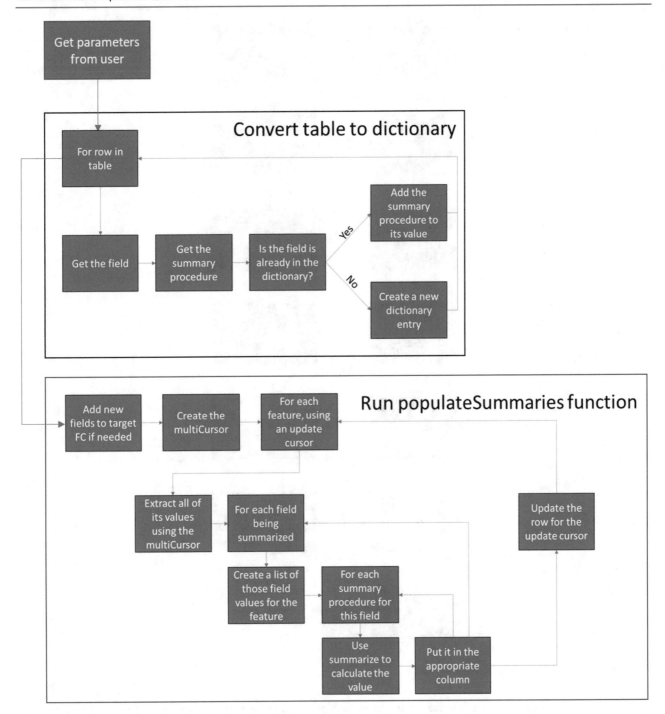

Fig. 20.9 Final flow chart for PValues_ch20.py

Fig. 20.10 Beginning flow chart for summarize.py

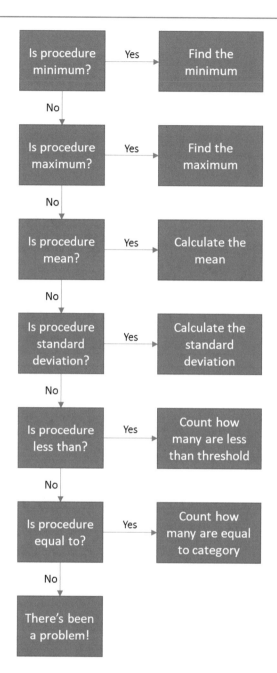

Fig. 20.11 Filling out the black box for summarize.py

py file that there was no guarantee that the order represented by the next method in this class and the order used by the `updateCursor` in that function would be the same order. This meant I needed a way to be able to get the values associated with a specified feature. Therefore, I had to go back and add the `extractFeature` method to the UML diagram (Fig. 20.15).

The next method, `extractField()`, is similar, in that it extracts a list of lists of values but keeping the same field instead of the same feature. This requires a more complex procedure, however, because search cursors are organized by feature, not by field. This means we have to put in a nested for loop to ensure the field and feature combinations are extracted in the right order. We have to first loop through the features, using the first feature class as an index. However, a local cursor must be created so we don't alter the index represented by the property holding a cursor of the first feature class. Looping through that property here would alter the index used by the `next()` method, which we don't want to do, so a different cursor for the same feature class is created here. This illustrates that while we can have methods call each other, we need to be careful that the methods do not have unintended side effects upon other methods, creating unexpected behaviors for the user or programmers employing this class. Then for each feature, we can create a series of cursors for the remaining feature

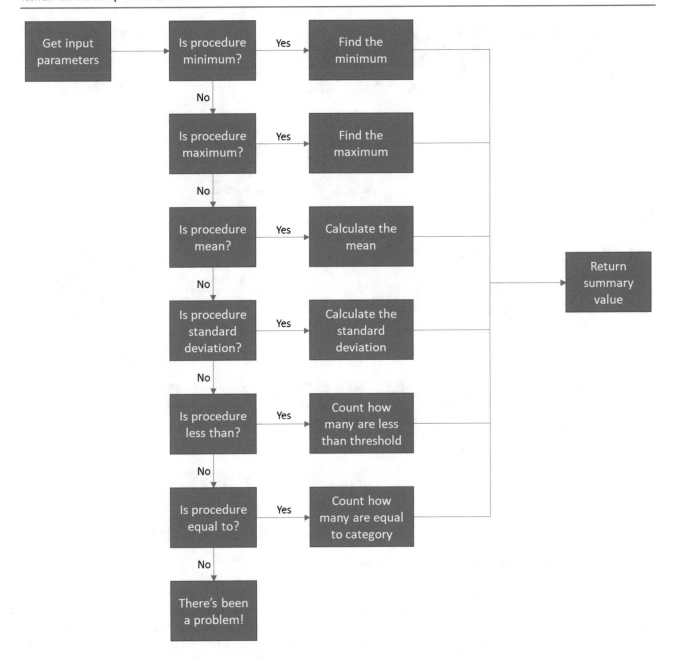

Fig. 20.12. Final flow chart for summarize.py

Fig. 20.13 Flow chart for multiCursor.py __init__() method

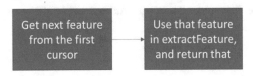

Fig. 20.14 Flow chart for multiCursor.py next() method

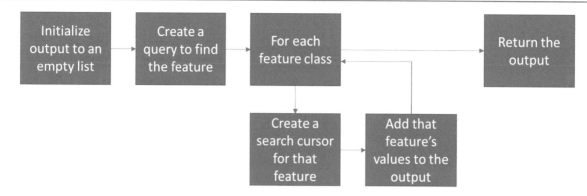

Fig. 20.15 Flow chart for multiCursor.py `extractFeature()` method

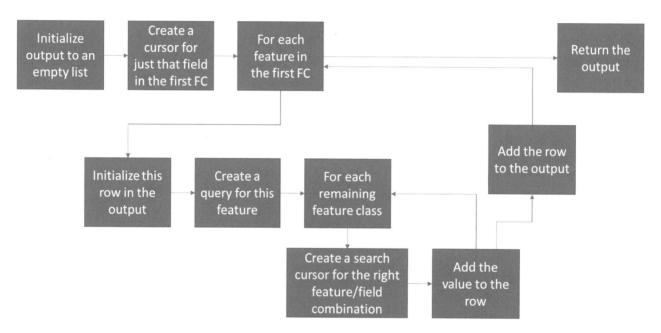

Fig. 20.16 Flow chart for multiCursor.py `extractField()` method

classes, finding the appropriate feature/field combination. This will then allow us to get the right value that needs to get appended to the output row. After all feature classes have been examined, that row can be added to the output list of lists. Then after all combinations of features and fields have been looped through, we can return the completed output (Fig. 20.16).

The last method is simple enough that it doesn't even need a flow chart. Since the next method uses the property of the cursor looping through the first feature class, we can reset the multiCursor by simply resetting that cursor.

Now that we have planned out all of the functions and methods, we can move to creating the code for the entire program. They will be taken in order of complexity.

Task 3: Write the Python Code for summarize.py

The simplest of the three files is the summarize file. The new module that is imported is a statistics module to calculate the standard deviation. The boxes in the flow chart have been turned into comments here, and we will fill in below the comments:

```
import statistics as STATS


#define the summarize function
```

```
#initialize variables


#if the procedure is the minimum


#elif the procedure is the maximum


#elif the procedure is the mean


#elif the procedure is the standard deviation


#elif the procedure is less than


#elif the procedure is equal to


#else it fit nothing


#return the output value
```

The first new material is how to specify optional parameters when defining a Python function. In the `def` statement, we can provide default values for input parameters by setting them equal to specified values. In the lines below, I set them up to be 0.05 for the threshold and "HH" for the category. Once there are default values, they are considered optional because even if the call to the function does not specify what they are, the computer can access a value for them in this function:

```
import statistics as STATS


#define the summarize function
def summarize(procedure, valueList, threshold=0.05, cat='HH'):
```

In the initialization step, we need to first set up the output variable. It is initialized to a value that will not be needed here, but we nonetheless had to set up the variable here. You will see that in every single `if` or `elif` clause, we will overwrite this value of -999. However, because there is an `else` clause that does not overwrite the value, we do need to have something here to ensure that when the computer tries to return the output value, there is a variable that exists to return. The other part of the initialization step is sorting the value list. The reason for this is for you to identify in Question 1, at the end of this task:

```
#initialize variables
    output = -999
    valueList.sort()
```

If the procedure is the minimum, we can simply now get the value from index #0 and make it the output. Likewise, we can use the last item in the list for the output if we want the maximum:

```
#if the procedure is the minimum
    if procedure == "MINIMUM":
        output = valueList[0]
```

```
#elif the procedure is the maximum
    elif procedure == "MAXIMUM":
        output = valueList[-1]
```

If the procedure is the mean, we can use built-in functions of `len` and `sum` to get the number of items in the list and their sum. Dividing the sum by the number of items gives us the mean, and we can easily return that:

```
#elif the procedure is the mean
    elif procedure == "MEAN":
        count = len(valueList)
        total = sum(valueList)
        output = total/count
```

If the procedure is the standard deviation, it would be possible to calculate that. However, when there is a module that includes a function to calculate it for us, why not use that? Therefore, we will use the `STATS.stdev()` function for this option:

```
#elif the procedure is the standard deviation
    elif procedure == "ST_DEV":
        output = STATS.stdev(valueList)
```

If the procedure is to count the number of items less than a certain value, we will need to loop through the list of values to get this. However, to make it a bit faster, we can use the fact that we have already sorted the list. If we start at the beginning and keep counting until we find a value that exceeds the threshold, the index we are at when we finish counting gives us the desired answer. The only catch is that with a while loop here, to avoid errors if all values are less than the threshold, we need to add another condition that the index is before the end of the list:

```
#elif the procedure is less than
    elif procedure == "COUNT_UNDER":
        soFar = valueList[0]
        index = 0
        while (soFar < threshold and index < len(valueList)):
            index = index + 1
            soFar = valueList[index]
        output = index
```

If the procedure is to count the number of items equal to a specified value, there's no such means to speed up the process. We have to investigate every item in the list and add to the output count each time that value is found:

```
#elif the procedure is equal to
    elif procedure == "COUNT_CATEGORY":
        output = 0
        for value in valueList:
            if value == cat:
                output = output + 1
```

Lastly, if nothing worked, then the procedure that was given must not have been a valid entry. In that case, it is best to raise a `ValueError` to provide a helpful error message to the user or the programmer. At the very end of the function is the expected return statement:

```
#else it fit nothing
    else:
        raise ValueError("Invalid procedure type: " + procedure)
```

```
#return the output value
    return output
```

You may have had a couple questions about programming decisions as you looked through this, and these are the first two questions.

Question 1: What was the purpose of sorting the list? Note: the ability to make finding the less than count was a nice side benefit, so there are other, more direct, benefits, too. Identify those more direct benefits.

Question 2: For the count of the values less than a threshold, our last index points to the first item *greater* than the threshold, but we did not subtract one to get the number of items less than the threshold. Why is it correct to *not* subtract one from the index?

Task 4: Write the Python Code for multiCursor.py

The next most complex file is the multiCursor file, even though it has several methods within it. Because it uses cursors and other `arcpy` functions, we need to import `arcpy` at the start. The other elements of the flow charts are listed below as comments, which we will fill in:

```python
import arcpy as arcpy


#define the class


#define the __init__ method


#create properties from the input parameters


#initialize other properties for a cursor of the first feature class and the
index of the key field

#define the next method


#move the primary cursor to its next feature


#extract that feature's values and return it


#define the extractFeature method with the fid as an input parameter


#initialize the output as an empty list


#create a query to find the feature that was provided


#for each feature class
```

```
#create a cursor for that feature


#get the feature's values from that cursor and add it to the output.


#return the output


#define the extractField method with the field as an input parameter


#initialize the output to an empty list


#create a cursor for just that field in the first feature class


#for each feature in that cursor


#initialize the row in the output to a list containing the value from that
feature/field combination

#create a query for this feature


#for each remaining feature class


#create a search cursor for that feature/field combination


#add the value to the end of the row


#when all feature classes are done, add this row to the output


#return the output


#define the reset function as resetting the primary cursor property
```

Defining the class is as simple as using the `class` keyword with the appropriate syntax:

```
import arcpy as arcpy


#define the class
class multiCursor():
```

Next we define the __init__() method, which is the constructor for objects of the class `multiCursor`. Recall that, as with any method, the first parameter must be self. The other input parameters are as specified earlier:

```
#define the __init__ method
 def __init__(self, fcList, fieldList, keyID):
```

We then create properties from these parameters that will ensure they are accessible throughout the code for this class. The properties all start with `self.` as part of their syntax:

```
#create properties from the input parameters
 self.fields = fieldList
 self.keyID = keyID
 self.featureClasses = fcList
```

The other two properties are initialized as a search cursor accessing the first feature class and using the field list provided as an input parameter. At this point, we could use `fieldList` or `self.fields`. The index of the key field in the field list is also accessed and stored for future reference:

```
#initialize other properties for a cursor of the first feature class and the index of the key field.
 self.primaryCursor = arcpy.da.SearchCursor(fcList[0], fieldList)
 self.keyIDIndex = fieldList.index(keyID)
```

Our first regular method is the `next()` method. As it has no input parameters in the UML diagram, the only input here is the `self` that all methods must have. It starts simply enough by advancing the `primaryCursor` property to the next feature in that cursor:

```
#define the next method
 def next(self):

#move the primary cursor to its next feature
 primaryCursorRow = self.primaryCursor.next()
```

The process of extracting the feature's value is only slightly more complicated, because the row returned by that advancement step has all the fields for the feature in question. However, we only need the key value to make sure we have the right information for the `extractFeature` method. This is why we stored the `keyIDIndex` property. It will tell us where in the list stored in this row variable to find the right value:

```
#extract that feature's values and return it
 keyIDValue = primaryCursorRow[self.keyIDIndex]
 output = extractFeature(keyIDValue)
```

That leads us to the `extractFeature` method. It starts easily enough by initializing the output to an empty list and then creating a query to find the feature. Recall that the `keyID` property stores the field used as the key field, so a query that searches for all features with the key value equal to the specified fid value will get us to the right feature:

```
#define the extractFeature method with the fid as an input parameter
 def extractFeature(self, fid):
#initialize the output as an empty list
 output = []
#create a query to find the feature that was provided
 keyIDQuery = '"' + self.keyID + '" = ' + str(fid)
```

Next, we loop through the feature classes and use that query to ensure we have a search cursor that is specific to the feature we are interested in:

```
#for each feature class
 for fc in self.featureClasses:
#create a cursor for that feature
 search = arcpy.da.SearchCursor(fc, self.fields, keyIDQuery)
```

The next step of getting the feature's rows is placed within a try block because it is possible that the feature class does not contain the feature in question. We try to get as the next row in the search cursor and append it to the output. However, if that feature isn't included in the feature class, we will get a StopIteration error. Because it might not be obvious to a user of this class that a stop iteration error corresponds to a missing feature, we catch that and instead raise a ValueError with a more helpful error message:

```
#get the feature's values from that cursor and add it to the output.
        try:
            row = search.next()
            output.append(row)
        except StopIteration as si:
            raise ValueError("The FID provided does not exist in at least one feature class in this
multicursor: " + fid)
```

When all this is done, we can return the output:

```
#return the output
        return output
```

Before moving on to the next method, here's another question regarding an algorithmic decision that might not be obvious. In the try block, we just get the next row in the search cursor, without confirming whether it was the right row, and we don't even try to get any other rows from the cursor.

Question 3: Why can we be certain search.next() will give us the right row and only the right row, if that feature ID exists within the feature class.

Next, we set up the extractField method with the field as its input parameter and again initialize the output to an empty list:

```
#define the extractField method
    def extractField(self, field):
#initialize the output to an empty list
        output = []
```

We need to create a cursor for just this field in the first feature class and, as stated above, must use a different cursor from the one stored in the property. Since we have to create a new cursor, we may as well narrow it down to just the key field and the field being investigated. This is within an if statement that checks to confirm that the field we are looking at exists within the field list. At the end, you will see the else clause for this, which raises a ValueError to inform the user or programmer that the desired field wasn't in the list for the multiCursor:

```
#create a cursor for just that field in the first feature class
    if field in self.fields:
        localPrimarySearch =
arcpy.da.SearchCursor(self.featureClasses[0], [self.keyID, field])
```

Once that is created, we loop through it and can initialize that feature's row in the output table with what will be its first element: the value from the first feature class. That is guaranteed to be feature[1] because of how we defined the search cursor above. We know that the relevant ID value is in item zero and the correct field value is in item one:

```
#for each feature in that cursor
        for feature in localPrimarySearch:
#initialize the row in the output to a list containing the value from that
feature/field combination
            thisRow = [feature[1]]
```

That the ID value is in item zero helps us create the query for the feature, so we can ensure that as we look at all the other feature classes, we are looking at the right feature for it. Because even if all the features are the same, there is no guarantee that the cursors will be in the same order, we have to individually search for each feature in the remaining feature classes to ensure that all values are properly aligned. Therefore, we need this query:

```
#create a query for this feature
            currentKey = feature[0]
            currentKeyQuery = '"' + self.keyID + '" = ' + str(currentKey)
```

We can then loop through all the remaining feature classes. Note that I am subsetting the list of feature classes by excluding the first one. That's the one we have already looked at and added to the row by initializing the row to its value, so this for loop can start at the feature class at index 1 instead of index 0:

```
#for each remaining feature class
            for fc in self.featureClasses[1:]:
```

We can use that query to generate a search cursor for the feature class. Within this cursor, we can use the same rationale as in Question 3 to simply get the next row, and we already know we are interested in item #1 within that row, based upon how the cursor was constructed. That value gets appended to the end of the row:

```
#create a search cursor for that feature/field combination
                search = arcpy.da.SearchCursor(fc, [self.keyID, field],
currentKeyQuery)
#add the value to the end of the row
                thisRow.append(search.next()[1])
```

At the conclusion of the row, add it to the output. Be careful about indentation here!

```
#when all feature classes are done, add this row to the output
            output.append(thisRow)
        else:
            raise ValueError("The provided field is not in the list of fields
for this multicursor: " + field)
```

Finally, we can return the output:

```
#return the output
        return output
```

The last method for this class is the simplest. Resetting the multiCursor resets the primary cursor so that if or when the next() method is called again, it will start over at the beginning:

```
#define the reset function as resetting the primary cursor property
    def reset(self):
        self.primaryCursor.reset()
```

This concludes the file, since it only has the class.

Task 5: Write the Python Code for PValue_ch20.py

This file contains the most complex logic, as represented by the most complex flow chart. That's why this one is last. As we go through, you might notice that the flow chart was designed to get input parameters from an ArcGIS tool, although, as alluded to in Task 1, we are sticking with IDLE here. That means that the approach to getting data in is different. Because this is ultimately a preliminary setup in preparation for connecting to a tool in Chap. 21, I have hard-coded the values into the file instead of using the `input()` function to get them from the user in the IDLE Python shell. If we were sticking with the IDLE shell, that function would be a better approach, although it is not nearly as flexible as ArcGIS Pro in allowing the user to easily input a list of feature classes or a value table of the field/summary procedure pairs. I will start with this code, even though it appears at the end of the file.

The first line sets up the workspace. The next four lines create the list of feature classes that will be used. All of them share a very similar name, with `LISA_Queen_Rand#_GA` in which only the number ranges from 1 to 10. After this list is set up, the target feature collection and the key field are defined. Lastly, instead of working through the value table as will be done later, the dictionary is directly created, using the field called `LMiPValue` and finding the minimum, maximum, mean, standard deviation, and counts of the number of values less than 0.1, 0.05, and 0.01. With all the input parameters thereby specified directly, we can call the `populateSummaries()` function:

```
ARCPY.env.workspace = "C:\\Users\\jfconley\\Documents\\ArcGIS\\Projects\\PValue\\PValue.gdb"
base = "LISA_Queen_Rand"
fcList = []
for i in range(10):
  fcList.append(base + str(i+1) + "_GA")
targetFC = "TrumpAndRandom_GA"
keyField = "SOURCE_ID"
fieldSumDict = {}
fieldSumDict["LMiPValue"] = ["MINIMUM", "MAXIMUM", "MEAN", "ST_DEV", "<0.1", "<0.05", "<0.01"]

 #populate the summaries
populateSummaries(fcList, fieldSumDict, targetFC, keyField)
```

Returning to the start of the file, we need to be able to ensure that this Python file is able to access the other two Python files. This is quite easy if and only if all files are stored in the same folder, which is what we will do here. We can use the same `import` statement to access these files we created. These are followed by the comments for each of the elements within the flow chart for this function:

```
import arcpy as ARCPY
import multiCursor as MULTICURSOR
import summarize as SUMMARIZE

#define the function and get the list of fields


#Add new fields to target feature class


#Create the multicursor


#for each feature in an update cursor of the target feature class


#extract that feature's values using the multicursor.
```

```
#for each field in the list of keys


#create a list of field values for this feature, one per feature class


#compile all the summary values for this field


#Put the summary value in the correct column in the output row


#use update cursor to populate the new fields in targetFC
```

Now we will fill in after each of the comments. It starts with defining the function using the input parameters taken from the UML diagram and getting the list of fields from the keys of the dictionary:

```
#define the function and get the list of fields
def populateSummaries(fcList, fieldSumDict, targetFC, keyField):
 fieldList = list(fieldSumDict.keys())
```

We cannot just add these fields to the target feature class and call it done, because there can be many summaries for each of these fields, and we should have a new field in the target class for each field/summary combination. That means creating a new list of all the fields that will be added to the target feature class. It is initialized as the key ID field, because we will need this later. Then, the remainder of this list is generated by looping through the fields in the `fieldList` and creating a new field for each of the summary procedures for each field. There is an `if/elif` statement in here because while "minimum" and "mean" are acceptable for a field name, the "<" and "." characters that can appear in a less than procedure string, notated as "<0.05," for example, or the "=" in an equality procedure, such as "=HH," are not permitted within field names. The content of the `if/elif` statement replaces < with LT_ and any decimal points that may appear with an underscore. Likewise, an equals sign is replaced with is_. The one new syntax structure here is the condensed way of getting the list of field names. Instead of having a for loop stretch over many lines, `[field.name for field in ARCPY.ListFields(targetFC)]` tells the computer to create a list because it has square brackets at the outside, with the items in the list taken from the `field.name` property of each `field` in the list given by `ARCPY.ListFields(targetFC)`. We need to get this list because we will get an error message if we try to add a field to a feature class that already has the field. Therefore, this is needed for the `if/else` statement confirming whether the new field is in the feature class' existing fields. I use the `pass` statement here if it is in the existing fields because it is easier to read with the condition in the affirmative, even though I want to do something when it isn't true:

```
#Add new fields to target feature class
    targetFCFields = [keyField]
    for field in fieldList:
        sums = fieldSumDict[field]
        for summ in sums:
            toAppend = summ
            if summ.startswith("<"):
                toAppend = "LT_" + summ[1:].replace(".", "_")
            elif summ.startswith("="):
                toAppend = "is_" + summ[1:]

            newField = field + "_" + toAppend
            targetExistingFields = [field.name for field in ARCPY.ListFields(targetFC)]
            if newField in targetExistingFields:
                pass
            else:
```

```
          ARCPY.AddField_management(targetFC, newField, "DOUBLE")

     targetFCFields.append(newField)
```

Once the fields are established in the target feature class, we can move on to the rest of the function by creating the multiCursor. This involves giving the cursor a list of both the fields in the dictionary and the key field. However, it is necessary to copy the list first because the cursor's list needs to have the key field, while `fieldList` should not. A copy must be made using the `.copy()` method, not just a simple assignment operation, because the assignment operation simply has two variables referring to the same list structure in memory, so appending a value to one appends it to the other.[3] Once that new field is created, the multiCursor can be initialized, as well as the list of summary values that will be used to store the results of each of the summary calculations:

```
#Create the multicursor
    cursorFieldList = fieldList.copy()
    cursorFieldList.append(keyField)
    mc = MULTICURSOR.multiCursor(fcList, cursorFieldList, keyField)
    summaryValues = []
```

Now we can create an update cursor of the target feature class, using the fields we have already created and added if necessary. Then loop through each of the features within this cursor. Because we initialized the list of target fields to begin with the key value, we can now access that value is item 0 in the row to ensure we have the appropriate feature ID value when extracting all of that feature's values from the multiCursor:

```
#for each feature in an update cursor of the target feature class
    update = ARCPY.da.UpdateCursor(targetFC, targetFCFields)
    for row in update:
        rowID = row[0]
#extract that feature's values using the multicursor.
        thisFeature = mc.extractFeature(rowID)
```

After a couple more initializations, which are needed to keep track of where in the lists of fields and summary values we are, we can loop through the fields in the list of fields taken from the dictionary. We start the summary values index at 1 instead of at 0 because the purpose of this index is to track where in the update cursor's row we are, and the first item in that row was the key ID:

```
#for each field in the list of keys
        currentFieldIndex = 0
        currentSummaryIndex = 1
        for field in fieldList:
```

Now that we are looking at each feature/field combination, we need to generate a list of all the values for this combination. We can use a shortcut here because `thisFeature` is the extracted data for this feature from the `multiCursor`. It should be constructed as a list of lists such that each of the items in `thisFeature` refers to a list of the values for a specific feature class and that inner list has the values in a consistent order, with the desired item always at the position represented by `currentFieldIndex`. Therefore, looping through each of the lists within `thisFeature` and getting the item with the current field index should give us the list of all values for this feature/field combination:

```
#create a list of field values for this feature, one per feature class
            values = []
            for cursorRow in thisFeature:
```

---

[3]This is the second instance I have encountered in which Python uses anything resembling call by reference instead of call by value. This is the other exception referred to toward the end of Chapter 2, Task 1.

```
            values.append(cursorRow[currentFieldIndex])
        currentFieldIndex = currentFieldIndex + 1
```

Having gotten our list of values, we can now work through the summaries. This should be straightforward Python to read. We get a list of all the desired summary procedures as the value associated with the field's key and give the `summarize()` function the information it needs. Note that the optional parameters to this function are provided only when they are needed and that they are specified through naming the parameter and giving it the value with the assignment operator, such as `cat=category` to assign the variable category to the parameter `cat`:

```
#compile all the summary values for this field
        summaries = fieldSumDict[field]
        for summary in summaries:
            if summary.startswith("<"):
                procedure = "COUNT_UNDER"
                threshold = float(summary[1:])
                thisSummary = SUMMARIZE.summarize(procedure, values, threshold=threshold)

            elif summary.startswith("="):
                procedure == "COUNT_CATEGORY"
                category = summary[1:]
                thisSummary = SUMMARIZE.summarize(procedure, values, cat=category)
            else:
                procedure = summary
                thisSummary = SUMMARIZE.summarize(procedure, values)
```

The remainder of the function is also straightforward. The value is put into the correct slot in the row, and that index gets incremented. Lastly, when all the values for that feature have been created and placed into the row, the update cursor gets updated. At the conclusion of the function, after all loops are done, the update cursor is deleted to ensure there are no lingering locks blocking anyone else from using this file:

```
#Put the summary value in the correct column in the output row
            row[currentSummaryIndex] = thisSummary
            currentSummaryIndex = currentSummaryIndex + 1
#use update cursor to populate the new fields in targetFC
        update.updateRow(row)

    del update
```

If you have everything entered accurately, and run the code, it will take a while, but finish. If you had ArcGIS Pro open with the target feature class, you will need to close that feature class so that this program can update that feature class. Please do so and answer the next questions.

Question 4: What is the maximum p-value (LMiPValue_MAXUMUM) for Seminole County?
Question 5: How many of the simulated iterations had a p-value less than 0.01 (LMiPValue_LT_0_01) for Jasper County?

Task 6: Unguided Work

In the unguided work (Task 7) of Chap. 16, you created a new class called Neighborhood. Take the class definition from your Chap. 16, Task 7 file, and create a new neighborhood.py file to define that class. Likewise, take the Parcel class and place it in a parcel.py file. Then take the remainder of that file, name it chapter20_task6.py, and have it complete the same functionality but referring to the new neighborhood.py file instead of having the class inside its file. To ensure it runs in IDLE instead of ArcGIS Pro, use the `input ()` function to ask the user for the appropriate environment workspace and all of the parameters that were in the tool for Chap. 16, Task 7. Likewise, replace all the output messages, errors, and warnings with `print ()` statements.

Evaluation criteria are as follows:

- The code is correctly divided between three files, and chapter20_task6.py correctly refers to neighborhood.py and parcel. py.
- The code correctly solicits input from the user through the `input ()` function and correctly provides information using the `print ()` function.
- The output generated is correct.

Developing a Custom Python Package

21

Introductory Comments: Python Packages

With a program that spans multiple files, sharing becomes more streamlined when we compile those files into a single package. In addition, having a package will make it possible to access our program here through an ArcGIS Pro tool. Packages are very commonplace within software development, as a semi-standardized unit of how code gets shared. Even if the terminology may change, packages are used across many programming languages and environments and have been used for decades.

The primary advantages of packages are their portability and their structure, and these are interrelated. The portability is evident in how many packages are shared through various online code repositories, such as GitHub or PyPI, the Python Package Index. By having packages available in these central repositories, it becomes much easier both to share your code for a wider audience and to find code that others have shared.

This portability only works, however, when there is a common structure underlying the packages. After all, every package you have imported, `arcpy`, `numpy`, and others, is a collection of Python files that the computer has made available to you. You did not have to open those files, install them yourself, or do anything else with them, which makes the importation process much easier for the vast majority of programmers who would not have the skills necessary to implement those connections manually. This means we must rely on the computer to implement those connections. Trusting the computer to make the connections properly means there must be a structure that is strictly adhered to in order to ensure the computer can find what it is looking for. This structure is examined and implemented in Task 1.

The remainder of the chapter takes the package that is created in Task 1 and enables it to be imported into the code for an ArcGIS Pro tool, which can then be packaged and shared as a .ppkx file.

Task 1: Turning the MultiCursor and Summarize Files into a Formal Package

The first aspect of sharing our code with the wider world and making it accessible to ArcGIS Pro is turning it into a package. In one sense, a package is simply a collection of files. However, while the computer is great at storing collections of files, if we want the computer to use those files without human supervision, they must be structured just as strictly as our computer code is. Any deviation from the structure that is expected of python packages will cause problems much like a syntax error prevents python code from being read.

To set up that structure, we will want to use a completely new folder and only have the minimum necessary inside that folder. Go to Windows File Explorer, and navigate to whatever file location you want to use for this chapter's work. Since this is continuing the PValue work from Chap. 20, I am using the same directory that I was using for that work. In your chosen directory, create a new folder, and call it "chapter21." Within that folder, create another folder, also called "chapter21." This gives the folder structure below, although the .idlerc folder will be automatically generated later, so you won't have it yet.

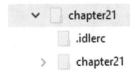

J. Conley, *A Geographer's Guide to Computing Fundamentals*, Springer Textbooks in Earth Sciences, Geography and Environment, https://doi.org/10.1007/978-3-031-08498-0_21

Within the inner chapter21 folder, place copies of the multiCursor.py and summarize.py files. If everything from Chap. 20 worked out correctly, you should have no need to edit these files. Simply copy them and paste them into the chapter21/chapter21 folder. These therefore form the basis of the package by providing the substantive content. However, the mere existence of these files in this folder does not provide the structural syntax that is needed for the package to work correctly. As an analogy, think of the pages of a book. These two files form the pages of the book. They give the information that is absolutely necessary for the book to have any instructive or entertainment value. However, if all you have are the pages without any binding, it all falls apart and cannot be actually read as a book. We need the binding. The rest of the package provides this binding to make sure the contents are usable. The binding takes the form of two additional .py files.

The first of these is much like the initialization method for a class and is named __init__.py. As with that method, it has two underscores before and after the init. To create this file, I suggest opening up IDLE and going to the File menu and then New file. Once it is opened, save that new file as __init__.py in the chapter21/chapter21 directory, so that it is sitting alongside summarize.py and multiCursor.py. After you have saved it, you are done. Oddly enough, this file is, for a project as simple as this one, empty. There can be other instructions here for more complicated packages, but nothing is needed for us. Nonetheless, even though it is empty, it is necessary to alert the Python system that this is intended to be a package.

The other file we need is up one directory, in the first chapter21 folder, alongside the inner chapter21 folder. It is to be called setup.py, and it has information about the package. There is also a strict structure for this file. It gives the metadata about the package through a `setup()` function, which should look very much like this. The names of the input parameters for the `setup()` function are mostly self-explanatory. The ones that might not be as readily apparent are the last three. The packages parameter uses the `find_packages()` function to extract everything from the chapter21 folder, as specified in the where parameter to the `find_packages()` function. The license parameter specifies a usage license that you want to use for your package. The MIT license is a commonly used one which says that it is free to use with the disclaimer that provided "as is" and thus the author is not legally liable for any consequences of bugs. You can use https://choosealicense.com/ to investigate your options. The last parameter specifies the packages that your package requires. In this case, because we import `arcpy` and `statistics`, these should also be available:

```
from setuptools import setup, find_packages


setup(
 name="chapter21",
 version="0.1",
 author="Jamison Conley",
 author_email="<jamison.conley@mail.wvu.edu>",
 description="Chapter 21",
long_description="Multicursor and summarize package for Chapter 21 of Geographer's Guide to
Programming",
 packages=find_packages(where='chapter21'),
 license='MIT license',
 install_requires=['arcpy', 'statistics']

)
```

Before proceeding any further, you should have the following folder and file structure. If it looks like this, you are ready to go (Fig. 21.1).

Installing packages is not done through IDLE. There are separate systems for this, one of which is called pip. To access it, we need to get to the command line prompt for Windows. I suggest using the search function next to the Windows start menu and searching for "cmd." You will then see "Command Prompt" as an app. Click on that, and you'll see this window appear.

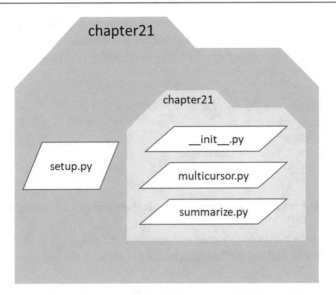

Fig. 21.1 Folder structure for the Python package

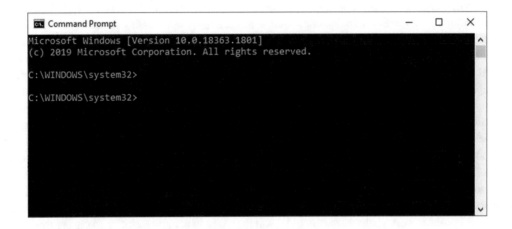

Type in "pip install chapter21" and press enter. Remarkably, you don't even need to navigate to the folder in which the chapter21 module structure is found. The pip program seems able to search the operating system and all associated files and find it itself. You might get output, but don't panic if you do not.

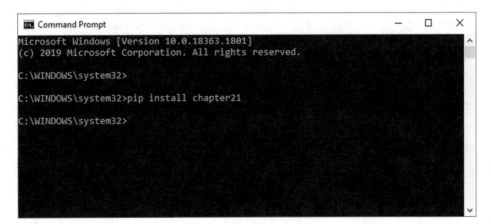

As long as there are no error messages, the installation was probably successful. Now we can check to ensure it worked by opening up a Python shell through IDLE. To do this, go to IDLE and click on the Run menu. Choose "Python Shell." Type in the following lines. Note that we need to specify which file we want from the chapter21 module. Then, we can access the summarize function within it:

```
>>> numbers = [1, 2, 3, 4, 5, 1, 2, 3, 4, 1, 2, 3, 1, 2]
>>> from chapter21 import summarize
>>> summarize.summarize("MEAN", numbers)
2.4285714285714284
```

Question 1: Using the imported summarize function, what is the standard deviation of this set of numbers?

Even though we are now able to access our module within a generic IDLE shell, ArcGIS Pro still cannot find it. We have to explicitly tell ArcGIS Pro how to find our module, and that is the focus of the next task.

Task 2: Telling ArcGIS Pro to Access the Package

A Python Script Tool, as we have previously been using, is great for creating a tool that runs a single script. However, as its name implies, it only handles one script.[1] The more flexible approach for multiple files is to use a Python Toolbox instead.

Following the guidance in the online help documentation for "Creating geoprocessing modules" from ESRI,[2] there is, like with Task 1, a strict structure to how the files are organized for a Python Toolbox. We will first get the structure set up and then create the toolbox to go into the structure. It will initially look like the structure from Task 1, although we will be enhancing this. At this stage, the only change you should make is to ensure that setup.py has a name, description, and long description appropriate for Chap. 21 instead of Chap. 20. The name must be chapter21 to match the folder next to it (Fig. 21.2).

The help documentation then suggests adding a series of additional folders, which will remain empty for now. Toolbox files may later go into these for distribution purposes, but this chapter is focused on debugging purposes and getting it into ArcGIS Pro especially, so this chapter does not concern itself with the contents of these folders. For now, it is better to have a structure that more closely mimics that from Task 1. We will have these folders, but leave them alone for now (Fig. 21.3).

In the course of testing the code, I found one area for improvement within the summarize.py file. To show where the new code fits into the `summarize()` function, the entire function is shown below with the new lines underlined. It turns out that the data I am analyzing has some <null> values in it, and Python is, understandably, unable to sort a list which contains the value `None`. Asking Python to sort a list which cannot be sorted results in a `TypeError`. To ensure this doesn't derail the entire analytical process, I added a `try/except` statement which automatically returns -999 if the data cannot be sorted. The use of a number like -999 is a common placeholder for missing data in GIS databases.

I am presenting this now instead of showing you the error and having you fix it because fixing code that is in the imported files and having that appear in the tool within the Python Toolbox are not as simple as making a change, clicking "save," and rerunning the tool. Each time you revise the associated files (summarize.py and multiCursor.py here), you must close ArcGIS Pro and restart it for the changes to be recognized by the tool. It is as if ArcGIS Pro compiles the imported files at the time of import and only bothers to look at whether they changed when the project is being assembled when it is loaded into ArcGIS Pro:

```
def summarize(procedure, valueList, threshold=0.05, cat='HH'):
    output = -999
    try:
```

---

[1]By placing Python files in the same directory, I was able to get it to access additional files, as in Task 1, but any corrections or changes to the imported files were never reflected in the compiled code. It appears ArcGIS Pro compiled the code into binary files that it stored somewhere and then never allowed for them to be changed. This led to very unusual error messages in which the error message reflected the old code, but the line printed on the screen was the new code that attempted to resolve the error message. As such, I recommend the Toolbox approach here unless you are *100%* confident the imported files will never need to be changed.

[2]https://pro.arcgis.com/en/pro-app/latest/arcpy/geoprocessing_and_python/extending-geoprocessing-through-python-modules.htm at the time of writing

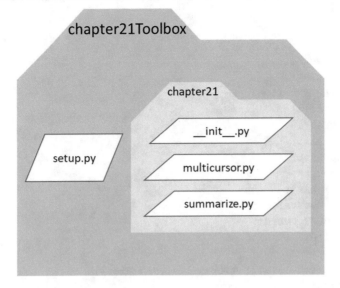

Fig. 21.2 Beginning folder structure for Python Toolbox

Fig. 21.3 Adding esri folder to structure for Python Toolbox

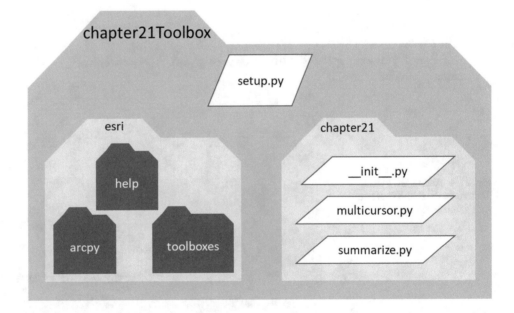

```
        valueList.sort()
except TypeError:
    return output
if procedure == "MINIMUM":
    output = valueList[0]
elif procedure == "MAXIMUM":
    output = valueList[-1]
elif procedure == "MEAN":
    count = len(valueList)
    total = sum(valueList)
    output = total/count
elif procedure == "ST_DEV":
    output = STATS.stdev(valueList)
elif procedure == "COUNT_UNDER":
    soFar = valueList[0]
    index = 0
    while (soFar < threshold and index < len(valueList)):
        index = index + 1
        soFar = valueList[index]
    output = index
elif procedure == "COUNT_CATEGORY":
    output = 0
    for value in valueList:
        if value == cat:
            output = output + 1
else:
    raise ValueError("Invalid procedure type: " + procedure)
return output
```

Fig. 21.4 Adding the Python Toolbox to the folder structure for Python Toolbox

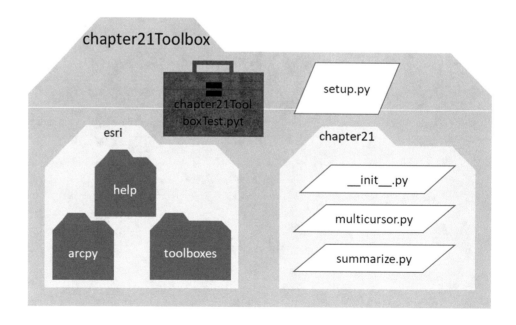

Question 1: Why would I immediately return the output in the `except` clause, instead of using `pass` to simply avoid sorting the unsortable list? In other words, think about what would happen if `pass` were used instead and why it is not the right choice here.

With this in place, we are now ready to create our Python Toolbox. This step is easy, with the only challenge being to ensure it is located in the desired folder. Go to the Catalog panel in ArcGIS Pro, and right-click on the Toolboxes entry. Select "New Python Toolbox." You will get a dialog box asking you where to put it and what name to use. The name is less important than the location as long as you are consistent with the name. The location must be in the chapter21Toolbox folder, alongside the chapter21 folder, setup.py, and the esri folder. As with the files in Chap. 20, the toolbox must be located here to ensure it can find the chapter21 package (Fig. 21.4).

This creates a toolbox with exactly one tool. That tool, at the moment, has no input parameters and does absolutely nothing. This is because the Python code associated with this toolbox is a placeholder file generated by ArcGIS Pro, and in the next task, we will need to revise it to make it do what we want, in this case, allow a tool to carry out the same functionality as in Chap. 20.

Task 3: Editing the Code to Work in an ArcGIS Pro Python Toolbox

You might have been wondering where the content of the PValue_ch20.py file was going to go, since only multicursor.py and summarize.py were copied into the chapter21 folder. This code will go directly into the Toolbox itself. This means editing the .pyt file for the toolbox. Don't panic at the change in extension. ArcGIS Pro uses .pyt for a Python Toolbox, and its contents are Python code. This means that it can be edited just like a .py Python file. Now that we are able to access the chapter21 package from within ArcGIS Pro, recall that we had to make slight adjustments to the PValue_ch20.py program to have the information from the ArcGIS Pro custom tool provided within the code instead of getting it from the tool. Because, unlike in Chap. 21, we will be running the code through our toolbox, we need to make the code appropriate for that setting. To access the toolbox code, simply right-click on the toolbox and click "Edit." This brings up a Python file that might look quite daunting at first, but let's take it in pieces and make the changes as we go along.

At the top are the imports. In addition to arcpy, we need our chapter21 imported code. This gives the following imports. Note that the MULTICURSOR and SUMMARIZE imports are capitalized so that we can reuse as much of the code as possible from Chap. 20 without changes:

```
import arcpy as ARCPY
import chapter21
from chapter21 import multiCursor as MULTICURSOR
from chapter21 import summarize as SUMMARIZE
```

Next is a custom class definition for the Toolbox. We will make minimal changes here, giving it more suitable values for the `.label` and `.alias` properties. I chose "ctt" for the alias as an acronym of Chap. 21 Toolbox. If we wanted more than one tool in this toolbox, we would define multiple tool classes after this (such as Tool1, Tool2, etc.) and have `self.tools` refer to a list of all the tools we define. Here, we only have one, so we can leave it as is:

```
class Toolbox(object):
    def __init__(self):
        """Define the toolbox (the name of the toolbox is the name of the .pyt file)."""
        self.label = "Toolbox"
        self.alias = "ctt"

        # List of tool classes associated with this toolbox
        self.tools = [Tool]
```

The remainder of this file is the tool itself, represented as a custom class. Let's take a look at each of its methods in turn. As is standard, it begins with the `__init__()` constructor method. In this, we only need to change the tool's `.label` property, which is the tool's name that will appear when you open the toolbox:

```
class Tool(object):
    def __init__(self):
        """Define the tool (tool name is the name of the class)."""
        self.label = "Chapter 21 toolbox"
        self.description = ""
        self.canRunInBackground = False
```

Next, we have the `getParameterInfo()` method. This one tells the toolbox what parameters to have for this tool and what properties they have, such as what data type they expect to receive from the user. We can revisit the UML diagram from the start of Chap. 20 to remind ourselves what these parameters ought to be. This will be helpful because we have to make all the specifications of the parameter setup through the code instead of through a GUI that creates the tool for us. Each of these will be examined in turn, as they are each represented by an object the `arcpy.Parameter` class, all of which are constructed within this method. Looking at the code below, it might not be immediately apparent that these are calls to the constructor for this class, because the input parameters to this constructor are presented vertically instead of all in the same row. As long as the input parameters are indented exactly one level, this is permitted. The commas at the end of all but the last line might give away that these are the input parameter tuples for a constructor:

- fcList: list of feature classes
- fieldTable: table of fields and strings
- keyField: field
- targetFC: feature class

The first is the list of input feature classes. Its display name and name properties reflect this purpose, and the difference is that the display name is what appears in the tool's dialog box, but the name is used internally by ArcGIS Pro and thus must adhere to variable naming requirements, like not containing spaces. There are plenty of data types available, and those are set through the datatype parameter, which is the `datatype` input parameter to the `__init__()` constructor of the `Parameter` class, to be precise. An examination of the online documentation for the arcpy `Parameter` class tells us that the desired value for a feature class is "DEFeatureClass." To tell the computer that we want this to accept a list of feature classes, not just one, we need to set the `multiValue` input parameter to `True`. We want to also ensure that this parameter is required and is an input parameter, as set by `parameterType` and `direction`, respectively. While there are other settings we can specify as input parameters to the `Parameter()` constructor, they are optional and not relevant here:

```
    def getParameterInfo(self):
        """Define parameter definitions"""
        param0 = arcpy.Parameter(
            displayName = "Input Feature Classes",
            name = "in_features",
            datatype = "DEFeatureClass",
            multiValue = True,
            parameterType = "Required",
            direction = "Input")
```

The second parameter for the tool that we are defining is the most complex, as it is a table. Much of it follows the same lead as our definition for `param0` above. The differences are that the names reflect this specific parameter, we do not need to specify that `multiValue` is false because that's the default for the `multiValue` parameter, and the datatype is different. Its datatype is `GPValueTable`, which is how arcpy represents this kind of input parameter. However, once we have created the `param1` object, we have to go back and specify the structure of this table. This is the last line, which fills in the columns property of the param1 object as a list of lists. Each of the internal lists has two elements: the data type for that column and its name. The outer list is a list of these elements. This means there are two columns to our list: a Field with the display name of "Field," and a string with the display name of "Summary Procedure":

```
        param1 = arcpy.Parameter(
            displayName = "Field/Summary Pairs",
```

```
                    name = "field_summary_pairs",
                    datatype = "GPValueTable",
                    parameterType = "Required",
                    direction = "Input")
                param1.columns = [["Field", "Field"], ["GPString", "Summary Procedure"]]
```

The remaining two parameters use ideas from the first two, so should be straightforward to understand:

```
            param2 = arcpy.Parameter(
                displayName = "Key Field",
                name = "key_field",
                datatype = "Field",
                parameterType = "Required",
                direction = "Input")

            param3 = arcpy.Parameter(
                displayName = "Target Feature Collections",
                name = "target_features",
                datatype = "DEFeatureClass",
                parameterType = "Required",
                direction = "Input")
```

With the four input parameters defined, we can now create a list of the parameters and return this list at the conclusion of the method:

```
params = [param0, param1, param2, param3]
return params
```

The next three methods do not need to be changed for the tool to work. For completeness' sake, I am presenting them here, although there are no edits to worry about:

```
    def isLicensed(self):
        """Set whether tool is licensed to execute."""
        return True

    def updateParameters(self, parameters):
        """Modify the values and properties of parameters before internal validation is performed.
This method is called whenever a parameter has been changed."""
        return

    def updateMessages(self, parameters):
        """Modify the messages created by internal validation for each tool parameter. This method
is called after internal validation."""
        return
```

Lastly, we have the execute() method. This is what happens when you click the "Run" button, so we definitely want to modify this to make it do something. Copying and pasting from the PValue_ch20.py file that we created is a good place to start, so that the tool works with that code. However, for some reason, Python would not recognize the populateSummaries() function by doing this, so the code from within the populateSummaries() function had to be copied and pasted into the space where that function had been called in PValue_ch20.py. This makes the execute() function very long, which isn't ideal, but seems to be the only option.

Recall that we had to change the code somewhat from the initial plan because we hard-coded the input parameter values. That is more properly done here, accessing the values that the user enters in as the parameters to the dialog box. Also note the start of a try/except block, designed to catch errors and ensure more user-friendly ArcGIS error messages are presented. The extensive use of .value properties might be surprising here, so allow me to explain. You might have found the text referring to the getParameterInfo() method confusing because we were referring to two types of parameters: one being the values that are inputs to the constructor method and the other being the Parameter object under construction. This means that if I get an item from the list of parameters below, it returns an object of type Parameter. This Parameter object refers to an item in the dialog box and contains information like its name, its data type, and its current value. To get that value, we need to start with the parameter's .value property. However, this does not immediately provide access to the value we are wanting, like the list of input feature classes, the key field, or the target feature collection. For the simpler data types, like the field and the target feature class, the .value property provides us an object of type Value. This might seem like overkill to have a class that basically serves as a simple container for the value of the property, but it is in contrast to the more complex data types. To get the actual value, or the content of that entry in the tool's dialog box, we need to get the .value property of the Value object. This is why parameters 2 and 3 have the seemingly redundant .value.value part to them. The first .value moves from the Parameter object to the Value object, while the second extracts the true value from the Value object.

For the more complex data types, like the list of feature classes, we have to work some to extract the desired information from that .value property. The first parameter is returned as a ValueTable object from the .value method. This is, of course, not the same thing as a list of feature classes, so we have to loop through each of the rows in this ValueTable and get the values from it one by one, to append them to a list that will become our feature class list. That's what the for loop below accomplishes:

```
def execute(self, parameters, messages):
    """The source code of the tool."""
    try:
    #get parameters from the tool: list of input FCs; list of fields; output target FC; key field

        fcVT = parameters[0].value
        fcList = []
        for val in range(fcVT.rowCount):
            fcList.append(fcVT.getValue(val, 0))
        fieldTable = parameters[1].value
        keyField = parameters[2].value.value
        targetFC = parameters[3].value.value
```

There is even more processing to convert the field table parameter into the dictionary we want. Perhaps confusingly, the parameter with the .datatype property of "GPValueTable" returned a list instead of a ValueTable. That means the exact code for looping through this parameter doesn't match what we used to create the list of feature classes, but the logic is similar. The for loop cycles through each row in the table. Recall from the way in which we set up the parameter that the first item in this is the field and the second is the procedure. To get the actual field, we need to, once again, use the .value property. However, this is not needed for the text string representing the procedure, stored in the proc variable. Instead of just adding this pair to the dictionary, however, we need to see if that field already exists as a key in the dictionary. If it does, that means we have seen this field before in the table and need to make sure we append the procedure to the list of summary procedures for this field instead of overwriting it. If it isn't in the list of keys, then we need to start a new list of procedures and add it as a new entry to the dictionary:

```
        fieldSumDict = {}
        for i in fieldTable:
            field = i[0].value
            proc = i[1]
            if field in fieldSumDict.keys():
                procedures = fieldSumDict.get(field)
                procedures.append(proc)
                fieldSumDict[field] = procedures
            else:
                procedures = [proc]
                fieldSumDict[field] = procedures
```

We have now extracted the information from the parameters in the way befitting a Python Toolbox instead of the IDLE shell we used in Chap. 20 or, for that matter, a Script Tool as in Chap. 14. At this point in time, the content of what had been the `populateSummaries()` function can be inserted without any changes. At the end, though, we have the `except` portion of the `try/except` block that was started at the beginning of the `execute()` function to replace Python error messages with more helpful ArcGIS error messages. The other difference is that the keyword return appears at the end to return nothing. This was in the `execute()` function to begin with, and I am opting not to remove it here:

```python
fieldList = list(fieldSumDict.keys())
#Add fields to targetFC
targetFCFields = [keyField]
for field in fieldList:
    sums = fieldSumDict[field]
    for summ in sums:
        toAppend = summ
        if summ.startswith("<"):
            toAppend = "LT_" + summ[1:].replace(".", "_")
        elif summ.startswith("="):
            toAppend = "is_" + summ[1:]

        newField = str(field) + "_" + toAppend
        targetExistingFields = [field.name for field in ARCPY.ListFields(targetFC)]
        if newField in targetExistingFields:
            pass
        else:
            ARCPY.AddField_management(targetFC, newField, "DOUBLE")

        targetFCFields.append(newField)

#Create the multicursor
cursorFieldList = fieldList.copy()
cursorFieldList.append(keyField)
mc = MULTICURSOR.multiCursor(fcList, cursorFieldList, keyField)
summaryValues = []

update = ARCPY.da.UpdateCursor(targetFC, targetFCFields)

for row in update:
    rowID = row[0]
    thisFeature = mc.extractFeature(rowID)
    #for each field in the list of keys
    currentFieldIndex = 0
    currentSummaryIndex = 1
    for field in fieldList:

        #extract the field from the multicursor
        values = []
        for cursorRow in thisFeature:
            values.append(cursorRow[currentFieldIndex])
        currentFieldIndex = currentFieldIndex + 1
```

```
                        #compile all the summary values for this field
                        summaries = fieldSumDict[field]
                        for summary in summaries:
                            #parse the < 0.05 and == 'HH' type
                            if summary.startswith("<"):
                                procedure = "COUNT_UNDER"
                                threshold = float(summary[1:])
                                thisSummary  =  SUMMARIZE.summarize(procedure,  values,  threshold=
threshold)

                                #carry out the summarization
                            elif summary.startswith("="):
                                procedure == "COUNT_CATEGORY"
                                category = summary[1:]
                                thisSummary = SUMMARIZE.summarize(procedure, values, cat=category)
                            else:
                                procedure = summary
                                thisSummary = SUMMARIZE.summarize(procedure, values)
                            row[currentSummaryIndex] = thisSummary
                            currentSummaryIndex = currentSummaryIndex + 1

                        #use update cursor to populate new fields in targetFC
                        update.updateRow(row)
                    del update

            except ValueError as ve:
                #catch any errors and convert to arcpy error messages
                ARCPY.AddError(ve)

            return
```

Since `execute()` is the last method of our tool, and we only have one tool in this toolbox, the code is complete, and we can now examine the tool.

Task 4: Examining the Tool

Open up the tool and give it a try. As you enter the feature classes, new rows will appear below the last one, indicating that you will get a list of feature classes instead of a single one. Also, you can see the columns for the Field/Summary Pairs Value Table, and, again, new rows will appear as you fill in existing ones.

Using the provided data, use the first ten LISA_Dist_Rand#_UT files to compute different summary statistics of the LMiPValue field. As in Chap. 20, the Key field is SOURCE_ID, and use TrumpAndRandom_UT as the output. Here is what it looks like.

After this provides the results, answer the following questions:

Question 2: How many of the first 10 simulations had p-values less than 0.01 for Salt Lake County?
Question 3: What is the standard deviation of p-values for the first 10 simulations for Tooele County?

Task 5: Unguided Work

Take the code from Chap. 20's unguided work, and create a new package, called neighborhood, for the neighborhood.py and parcel.py files. Place the code from the primary method into a new Python Toolbox that also accesses the neighborhood.py and parcel.py files. The tool within that toolbox can match the input parameters of your tool for Chap. 16, but should use the Toolbox script file for Chap. 21 here.

Grading criteria are as follows:

- The package is appropriately constructed, with the correct folder and file structure.
- The Toolbox's tool has the correct input parameterization in the `getParameterInfo()` method.
- The Toolbox's tool uses the new package, and correctly accesses it.
- The tool provides the correct behavior for inputs that are acceptable and are unacceptable based upon the development criteria laid out in Chap. 16.

Concluding Remarks

Having made it all the way to the end of the journey here, some perspective may be in order. You have taken steps from beginning to program, learning what variables, operators, loops, and conditional statements are, all the way to creating your own tools and toolboxes within ArcGIS. These tools have include several functions you have written yourself, as well as classes you created yourself, spread across several files. To go all the way from the start to the finish, here is an accomplishment, and for many of you, will provide the skills you need to begin and, hopefully, continue a career in GIS using or developing programs for a long time to come. Furthermore, I hope you have the skills and confidence to make the transition away from Python if and when ESRI changes programming languages. For that matter, these skills can help you move from the ESRI line of products to open source GIS programs like QGIS.

Even so, no single book can even attempt to cover the wide range of topics that could conceivably fit into the realm of GIS programming. Some of these may be useful resources for you if you seek to deepen your knowledge about one or more of the following topics.

GIS Algorithms

Perhaps the middle chapters sparked your interest in the nuts and bolts of the specific algorithms that enable the GIS to work, such as "how does the computer actually take the two geometries and find their intersection?" A couple resources for this can be the following books.

Wise, Stephen (2014). *GIS Fundamentals, 2nd ed*. Boca Raton, FL: CRC Press.

Xiao, Ningchuan (2016). *GIS Algorithms*. London: SAGE.

Gold, Christopher (2016). *Spatial Context: An Introduction to Fundamental Computer Algorithms for Spatial Analysis*. Boca Raton, FL: CRC Press.

Lovelace, Robin, Jakub Nowosad, & Jannes Muenchow (2019). *Geocomputation with R*. Boca Raton, FL: CRC Press.

Web GIS

Maybe sharing data and code has gotten you interested in working more with Web GIS. If so, I can suggest these resources.

Fu, Pinde (2020). *Getting to Know Web GIS, 4th ed*. Redlands, CA: Esri Press.

Dorman, Michael (2020). *Introduction to Web Mapping*. Boca Raton, FL: CRC Press.

Open Source GIS

Maybe after the end of this book, you feel confident enough to move away from ESRI entirely and you want to start working with or even contributing to open source GIS projects. If so, here are a couple places to start.

Neteler, Markus & Helena Mitasova (2013). *Open Source GIS: A GRASS GIS Approach*. New York: Springer.

Menke, Kurt, Richard Smith, Jr., Luigi Pirelli, & John Van Hoesen (2016). *Mastering QGIS, 2nd ed*. Birmingham, UK: Packt.

Whether you continue to more advanced programing or not, I hope you have gained valuable nformation and skills from this book, and wish you all the best in your future endeavors.

J. Conley, *A Geographer's Guide to Computing Fundamentals*, Springer Textbooks in Earth Sciences, Geography and Environment, https://doi.org/10.1007/978-3-031-08498-0

Appendix: Unified Modeling Language (UML)

As you move from simpler to more complex programming, from keeping track of simple variables, to objects, to classes, and ultimately managing projects containing multiple files, having a way to represent all the pieces of your code becomes important. A common way of managing this more complex code is through Unified Modeling Language, or UML. While UML is intended for managing software systems that are far more complex than what it presented within this book, a few aspects nonetheless can be useful here. There are 13 different types of diagrams in this visual language[1] within 3 broad categories. As such, an appendix here can only touch the surface of the complexity that UML can work with, so if you are interested in more information after reading this, I suggest looking at the UML organization's website at http://www.uml.org and following the references noted therein.

UML is a visual language which can represent a range of levels of abstraction of the processes inherent in constructing and maintaining a software system. The first category represents the structure diagrams, which depict how different objects, classes, and packages are constructed within the code. The UML diagrams presented within this book are from this category, so this is what is presented more fully within this appendix. The other two categories represent aspects of software development that tend to arise and become more relevant with more complex applications than we are building within this book. Briefly, behavior diagrams depict the processes that take place within the software system. Diagrams in this category are more focused on what the components of the system do, rather than how they are constructed. They can also represent the actions the users may want the software to do. The third type of diagram, interaction diagrams, depict how the different parts of the software system communicate with each other. As this book has focused primarily upon the structures within coding for GIS applications, this appendix focuses on the structure diagrams that are introduced in the chapters to provide a more detailed interpretation of how those diagrams are constructed and interpreted.

Class Diagram: A Single Class

Our entry point into UML is through the representation of a single class. Recall from Chap. 7 that classes contain properties, which are like object-specific variables and methods, which are like object-specific functions. The diagram for a single class concisely depicts these parts of a class.

Figure 7.1, reproduced here as Fig. A.1, introduced the concept of an object by extending the box metaphor for variables, by presenting an object as a crate which can hold a series of boxes. Thus, any object of the class `ArcGISProject` is guaranteed to have properties of the type `activeMap`, `dateSaved`, `defaultGDB`, etc., as well as methods called `save()` and `listLayouts()`, among others. While the colors of the boxes can represent the types of properties, such as green for text strings, this approach becomes difficult to scale to larger classes. UML class diagrams provide a more formal structure to this sort of visualization. Figure A.2 gives a UML class diagram version of Fig. A.1.

The class diagram here has three boxes, matching the three portions of the metaphorical crate. Like the label on the top of the crate, the top box contains the name of the class. The properties that are in the boxes in the left side of the crate correspond to the information in the middle of the three boxes in the class diagram. Each property has a single line. The name of the property comes first, followed by the type of that property after the colon. The methods, which are on the right side of the crate, are presented in the lowermost box in the class diagram. Two methods are presented. The name of the method

---

© The Editor(s) (if applicable) and The Author(s), under exclusive license to Springer Nature Switzerland AG 2022
J. Conley, *A Geographer's Guide to Computing Fundamentals*, Springer Textbooks in Earth Sciences,
Geography and Environment, https://doi.org/10.1007/978-3-031-08498-0

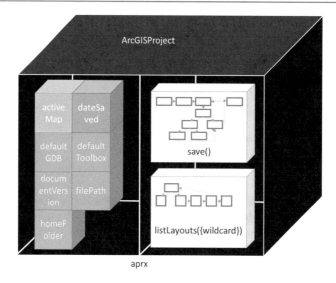

Fig. A.1 An object variable of the class `ArcGISProject`

ArcGISProject
activeMap: Map dateSaved: DateTime defaultGDB: String defaultToolbox: String documentVersion: String filePath: String homeFolder: String
save() : void listLayouts({wildcard : String}) : List

Fig. A.2 Class diagram representation of Fig. A.1

comes first, followed by the parameters. As in the code itself, if there are no parameters, a pair of parentheses containing nothing is still included. If there are parameters, then those are included within the parentheses as a name : type pair, as was done with the properties. Using the approach of the ESRI documentation, optional parameters are indicated here with curly braces. Lastly, the type of return value from the method is given after the final colon in the line. If nothing is returned, then the word "void" may be used to make this lack of return value explicit. Therefore, every `ArcGISProject` has a method called `save()` that takes no parameters and returns no values. Meanwhile, there is also a method called `listLayouts()` that has an optional wildcard parameter that is a string, and it returns a list to the code. While we know that this list must contain `Layout` objects, this is not explicitly guaranteed by Python, as lists can contain multiple types of entries, and, therefore, the return type here is simply denoted as a list.

One advantage of class diagrams here is that they are more compact than the crate approach. This means that it is feasible to represent larger classes. In fact, only a portion of `ArcGISProject` had been presented in the crate in Fig. A.1. Figure A.3, below, presents the entire class structure for `ArcGISProject` in a single diagram.

Class Diagram: Connecting Multiple Classes

Chapter 12 introduced another layer to the visual representation of class structures. When working with many classes at once, it can be very useful to depict how those classes relate to each other. An example of this is within the context of navigating the web of interconnected classes within the `arcpy.mp` module. For example, if you have an ArcGIS project and you want to change the color ramp used by a particular layer containing a choropleth map, it is not immediately obvious how to navigate from the `ArcGISProject` object to a `ColorRamp` object to edit this part of the layout. Using the diagrams from Figures 12.1 and 12.2, reprinted below as Figs. A.4 and A.5, and following the arrows in them, we can see the chain of

```
┌─────────────────────────────────────────────────────────────────────┐
│                            ArcGISProject                              │
├─────────────────────────────────────────────────────────────────────┤
│ activeMap: Map                                                        │
│ dateSaved: DateTime                                                   │
│ defaultGeodatabase: String                                            │
│ defaultToolbox: String                                                │
│ documentVersion: String                                               │
│ filePath: String                                                      │
│ homeFolder: String                                                    │
│ metadata: Metadata                                                    │
├─────────────────────────────────────────────────────────────────────┤
│ importDocument(document_path: String {include_layout : Boolean},      │
│                    {reuse_existing_maps : Boolean}) : void            │
│ listBrokenDataSources() : List                                        │
│ listColorRamps({wildcard : String}) : List                            │
│ listLayouts({wildcard : String}) : List                               │
│ listMaps({wildcard : String}) : List                                  │
│ listReports({wildcard : String}) : List                              │
│ save() : void                                                         │
│ saveACopy(file_name : String)                                         │
│ updateConnectionProperties(current_connection_info : String,          │
│                   new_connection_info : String,                       │
│                   {auto_update_joins_and_relates : Boolean},          │
│                   {validate : Boolean}, {ignore_case : Boolean}) : void│
└─────────────────────────────────────────────────────────────────────┘
```

Fig. A.3 Full class diagram representation of `ArcGISProject`

methods that would be needed to get from a project to that color ramp. We first get a `Map` through the `listMaps()` method of the `ArcGISProject` object. From the map, the `listLayers()` method is used to access a `Layer`. The `Layer` has a `Symbology` accessed through the `symbology` property. Likewise, the `renderer` property is used to access a `Renderer` object. A choropleth map suggests this `Renderer` would be of the `GraduatedColorsRenderer` subtype, which has a `colorRamp` property that can allow accessing, and thereby editing, of that color ramp. It should be noted that for clarity, as well as ensuring each figure fits on a single page, only a small subset of the properties and methods are included for each class.

The new elements of the diagrams here are the connections, represented by the arrows and associated numerical annotations. There are two types of arrows. The first has a diamond head. These indicate associations in which one class contains a reference to another, and the second class can be considered as part of the first. For example, each ArcGIS project contains one or more maps, as well as one or more layouts. Meanwhile, those `Map` and `Layout` objects do not generally exist outside of an `ArcGISProject` object. Likewise, a map may contain any number of tables and layers, yet in the code, those `Table` and `Layer` objects are only accessed via the `Map` object.

The other type of arrow has a standard triangular arrow head and represents superclass and subclass relationships, like those described in Chap. 8. Here, there are two such examples: a `RasterClassifyColorizer` and a `RasterUniqueValueColorizer` as subclasses of `Colorizer`, and the four types of renderer as subclasses of `Renderer`.[2] Because there are no generic colorizers or generic renderers, and everything must be one of the subclasses, these are considered abstract classes, which is indicated by italics.

The other item here are the numerical annotations, which indicate the cardinality of the relationships. Each association arrow has two numbers, which indicate how many of each class can be associated with the other. For example, an `ArcGISProject` can have any number of `Map` objects in its list, including 0. Likewise, it can have any number of `Layout` objects, including 0. Therefore, both of these have 0..n marked. However, each `Map` can be associated with only one `ArcGISProject`, and the same is true of each `Layout` object. By contrast, each `Layer` has exactly one `Symbology`, and each `Symbology` is associated with a single `Layer` through the symbology property. This puts a 1 at both ends of the arrow. Lastly, each `Symbology` has either a `Colorizer` if it contains raster data or a `Renderer` if it contains vector data. Since this means a `Symbology` has zero or one `Colorizer`, and zero or one `Renderer`, this arrow uses a 0..1 notation.

---

[2]Even this is not fully representative of the documentation. There are no Colorizer and Renderer classes mentioned in the documentation, even though the code seems to behave as if there are abstract classes for Colorizer and Renderer. This absence of an explicit abstract class is why the colorizer and renderer properties have a generic Object type instead of anything more specific.

Fig. A.4 Class diagram for a portion of the `arcpy.mp` module, focusing on raster symbology

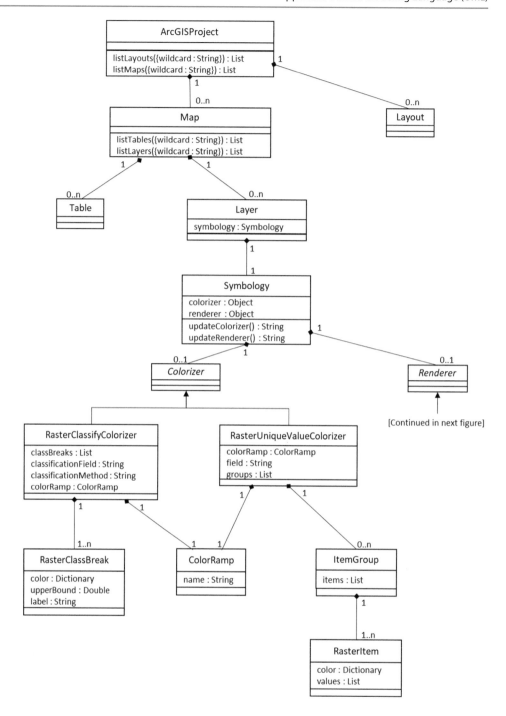

Object Diagram

While not included in the main text of the book, one other UML diagram can be useful for tracking down complex bugs: the object diagram. It looks very similar to a class diagram. The difference is that the class diagram shows the ways the different classes relate to each other in general, while an object diagram provides the specific values of the object variables at a point in time while the code is running. Therefore, to focus on the variables and their properties, a few changes get made. Firstly, the variable name gets included in the header. Secondly, the types of properties are replaced with their exact values. Thirdly,

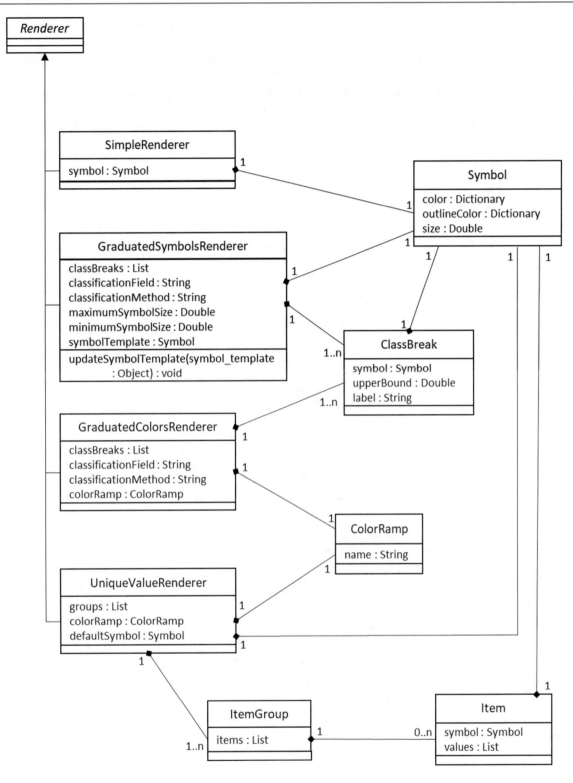

Fig. A.5 Class diagram for a portion of the `arcpy.mp` module, focusing on vector symbology

the methods section is removed, as it is no longer required. Lastly, parts of the diagram that are irrelevant, such as subclasses that are not currently in use, are removed.

As an example, Fig. A.6 represents a project containing a map with a single layer. In that layer is map using a simple renderer with dark gray symbols of size 10 points with black outlines.

This can help with debugging by providing a more intuitive visualization of how the different variables relate to each other than just looking at a list of variables in the lower portion of the IDLE debugger. For example, if the map in the project represented by the `aprx` variable looks wrong, perhaps because a typo left the size at 100 instead of 10, filling in the values of an object diagram can help you recognize both where the problem lies and how to fix it.

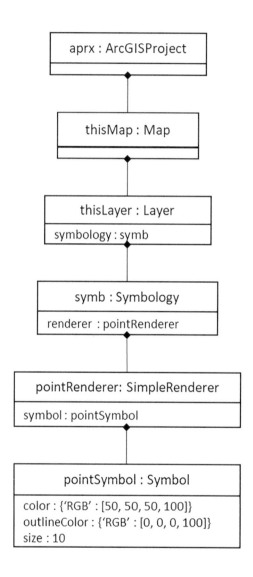

Fig. A.6 Example of an object diagram

References

Abler, R. F. (1988). The National Science Foundation National Center for Geographic Information and Analysis. *International Journal of Geographical Information Systems, 1*, 303–326.

Adams, D. (1987). *Dirk Gently's Holistic Detective Agency*. Pan.

Conley, J., Hong, I., Williams, A., Taylor, R., Gross, T., & Wilson, B. (2021). Assessing consistency among indices to measure socioeconomic barriers to health care access. *Health Services and Outcomes Research Methodology*. https://doi.org/10.1007/s10742-021-00257-5

Dorman, M. (2020). *Introduction to web mapping*. CRC Press.

Fu, P. (2020). *Getting to know web GIS* (4th ed.). Esri Press.

Gold, C. (2016). *Spatial context: An introduction to fundamental computer algorithms for spatial analysis*. CRC Press.

Lovelace, R., Nowosad, J., & Muenchow, J. (2019). *Geocomputation with R*. CRC Press.

Menke, K., Smith, R., Jr., Pirelli, L., & Van Hoesen, J. (2016). *Mastering QGIS* (2nd ed.). Packt.

Neteler, M., & Mitasova, H. (2013). *Open source GIS: A GRASS GIS approach*. Springer.

Ward, M. D., & Gleditsch, K. S. (2008). *Spatial regression models*. SAGE.

Wise, S. (2014). *GIS fundamentals* (2nd ed.). CRC Press.

Xiao, N. (2016). *GIS algorithms*. SAGE.

Index [1]

[1] Notes: items in the courier new font are Python elements. Keywords, such as in and or are simply the word, and are lowercase. Functions like len() and print() have parentheses after the function name to show they often expect parameters. Properties, such as .bandCount and .color begin with a . to indicate that they are associated with a variable which precedes the dot. Object methods, such as .index() and .keys() begin with a . to indicate their association with a variable and also have parentheses to show that they often expect parameters. The class associated with properties and methods is listed in brackets after the entry. Package names, like ARCPY and ARCPY.MP are in all-caps. Functions within packages, like reader() are presented with the package in brackets after the entry included so you know what to import. Many of these, but not all, are constructor methods, meaning they create an instance of the class by that name. Classes for which we did not encounter the constructor, like Map, are presented with their associated package, but without parentheses.

Printed in the United States
by Baker & Taylor Publisher Services